ThinkLab Algebra

ThinkLab Algebra

For Upper Primary Problem Solvers

Tamim Shahriar

Inside this book

A visual-first introduction to algebra. Uses bar models to make word problems concrete before introducing equations, then builds through expressions, solving, sequences, and competition-style thinking.

Tamim's ThinkLab

ThinkLab Algebra: For Upper Primary Problem Solvers

Published by Tamim's ThinkLab
Sydney, Australia

First edition, 2026

ISBN: 978-1-7646492-0-9

For bulk orders, school enquiries, or feedback, contact:
tamim@tamimsthinklab.com

To my amazing students at ThinkLab —
you ask the best questions.

Contents

How to Use This Book

A short guide for students

- Work through the chapters in order. Each chapter builds on the previous one.
- Show your working clearly, even when the arithmetic feels simple.
- Use the worked examples to learn the method, not just the answer.
- Try the practice questions on your own before checking the short answers.
- For harder questions, focus on the structure first: what is changing, what is equal, and what can be simplified?

A ThinkLab habit

When you get stuck, do not rush.
Ask:

- What do I know?
- What is the unknown?
- What equation or pattern connects them?
- Can I check my answer quickly?

A Note to Parents

This book is written for your child, but it works best when you are nearby.

A few things I would like you to know before they begin.

Go slow. One chapter at a time.

This is not a book to finish in a week. Each chapter introduces a way of thinking, not just a set of techniques. A good pace is one chapter every one to two weeks, with a small amount of practice on most days. Short and regular beats long and occasional.

Read it more than once.

The first reading might happen in the middle of Year 5, when your child is ready for something beyond the standard syllabus. The second reading could come in Year 6. The final reading should happen as they begin Year 7, when algebra becomes part of school. Each reading will feel different, and that is the point.

Let them struggle a little.

Please do not solve the problems for them. If they are stuck, ask a question instead:

- What do you know so far?
- What is the unknown?
- Can you draw a picture?
- Have we seen something like this before?

A small hint, given at the right moment, teaches more than a full solution.

Do not rush to algebra.

Algebra is powerful, but it is not always the fastest or most beautiful way to solve a problem. Before your child starts this book, they should have spent time on word problems using bar models, logical reasoning, and arithmetic thinking. Algebra should feel like a new tool added to a growing toolkit — not a shortcut that replaces everything else.

Many of the problems in this book can be solved without algebra. That is intentional. When your child compares two methods and sees why one is cleaner, they are learning something deeper than either method alone.

Celebrate the thinking, not the answer.

A wrong answer after careful thinking is more valuable than a right answer copied from somewhere. Ask your child to explain how they solved a problem. If they can explain it, they understand it.

Thank you for being part of their learning.

— Tamim Shahriar
Tamim's ThinkLab

Visual Algebra

Shapes, Balance, and Logic

Algebra starts with a single idea: a shape, a box, or a letter can stand for an unknown number. We start with shapes because they make the reasoning easy to see. Later chapters swap the shapes for letters. The thinking stays the same.

Three core moves in this chapter

As you work through these questions, keep these three moves in mind:

- **Balance:** the equals sign means both sides have the same value.
- **Substitute:** if you know a shape's value, replace it carefully.
- **Do the same to both sides:** this keeps the balance true.

The Main Idea

Concept 1: The equals sign means balance

In algebra, the equals sign does **not** just mean "here is the answer".
It means:

left side = right side

So both sides must have the same value.

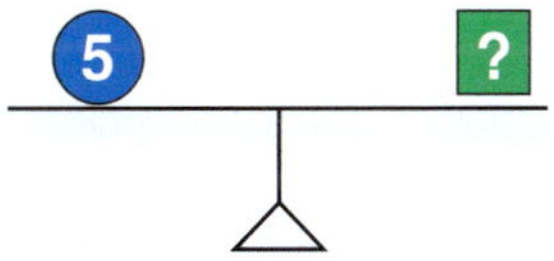

If the scale is balanced, the unknown box must also be worth **5**.

Rule: the same shape always stands for the same number

Within a single problem, identical shapes always represent the same number.
So if one ■ equals 5, then **every** ■ in that problem equals 5.
Different shapes **may** represent different numbers.
ThinkLab habit: before solving, check how many different shapes appear. Each distinct shape is a separate unknown.

Concept 2: Substitute carefully

If you know that a shape stands for a number, you can replace that shape with the number.
For example, if

$$\bigstar = 3,$$

then

$$\bigstar + \bigstar + 10 = 3 + 3 + 10 = 16.$$

ThinkLab habit: substitute first, then calculate step by step.

Concept 3: Do the same to both sides

If you add, subtract, multiply, or divide by the same amount on both sides of an equation, the equation stays true.
For example:

$$\blacksquare + \bullet = \bullet + 10$$

Remove ● from both sides:

$$\blacksquare = 10$$

Why does this work? Because both sides stay balanced.

Example of dividing both sides:

$$\blacksquare + \blacksquare + \blacksquare = 15$$

Divide both sides by 3:

$$\blacksquare = 5$$

The same logic applies to multiplication, addition, and subtraction.

Common trap: changing only one side

Be careful not to "cancel" something unless it appears on **both sides**.
For example:

$$\bigstar + 5 = 12$$

Some students rub out the 5 on the left and write

$$\bigstar = 12.$$

That is wrong. The left side became smaller by 5 while the right side stayed the same, so the two sides no longer balance.
Instead, subtract 5 from both sides:

$$\bigstar = 12 - 5 = 7$$

The key question is always: **am I doing the same thing to both sides?**

Worked Examples

Type A: Find the value of one shape

Example 1: The triangle mystery

Problem: Find the value of the triangle (▲) if

$$▲ + ▲ + ▲ = 12$$

Step 1: Rewrite the repeated addition

$$3 \times ▲ = 12$$

Step 2: Divide both sides by 3

$$▲ = 12 \div 3$$

$$▲ = 4$$

Quick check

$$4 + 4 + 4 = 12$$

Answer: ▲ = 4.

Type B: Use one clue to solve another

Example 2: Fruit balance logic

Problem:

- 2 apples balance 1 pineapple.
- 1 pineapple balances 4 bananas.

How many bananas weigh the same as 1 apple?

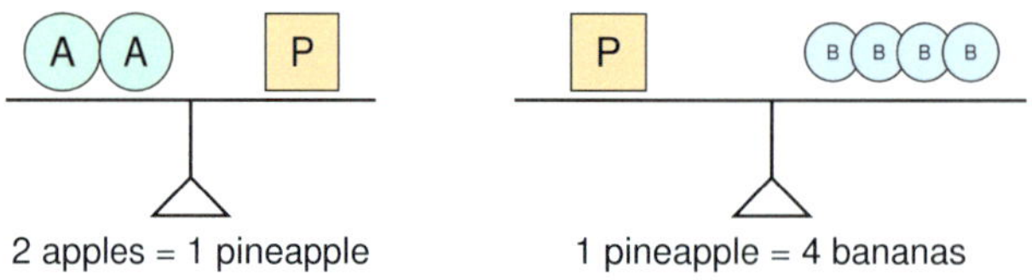

Step 1: Write the two facts

$$\text{Apple} + \text{Apple} = \text{Pineapple}$$

$$\text{Pineapple} = 4 \times \text{Banana}$$

Step 2: Substitute

$$\text{Apple} + \text{Apple} = 4 \times \text{Banana}$$

Step 3: Simplify Two apples equal 4 bananas, so one apple must equal 2 bananas:

$$\text{Apple} = 2 \times \text{Banana}$$

Answer: 1 apple = **2 bananas**.

Type C: Remove the same thing from both sides

Example 3: Clear the clutter

Problem: Find the value of the circle.

$$\bullet + \blacktriangle = \blacktriangle + 7$$

Step 1: Notice that ▲ appears on both sides
Subtract ▲ from both sides:

$$\bullet = 7$$

Answer: ● = 7.

Type D: Use one equation before the next

Example 4: The double clue

Problem:

$$■ + ■ = 10$$

$$■ + ★ = 9$$

Find the value of ★.

Step 1: Find the square

We can write ■ + ■ as 2■. (The 2 in front means “2 times”.)

$$2■ = 10$$

$$■ = 5$$

Step 2: Substitute into the second equation

$$5 + ★ = 9$$

Step 3: Solve

$$★ = 9 - 5$$

$$★ = 4$$

Answer: ★ = 4.

Before you practise

Quick self-check. Can you answer these without looking back?

- What does the equals sign mean in algebra?
- If the same shape appears twice in one equation, what must be true?
- What does it mean to “do the same thing to both sides”?

If you are unsure about any of these, re-read the concept boxes above.

Training Exercises

1. The double box

Find the value of ■:

$$■ + ■ = 18$$

2. The mixed sum

If ★ = 5, find the value of the heart (♡):

$$★ + ♡ = 12$$

3. The balance beam

What is the circle worth?

$$● + ■ = ■ + 11$$

Hint: Remove ■ from both sides.

4. Two equations, one unknown

$$■ + ■ = 16$$

$$■ + ★ = 13$$

Find the value of ★.

5. Fruit logic

- 3 oranges weigh the same as 1 melon.
- 1 melon weighs the same as 6 kiwis.

How many kiwis weigh the same as 1 orange?

6. The shape stack

■ + ■ + ● = 20

If ● = 8, what is ■?

7. The difference

▲ = ■ + 5

If ■ = 4, what is ▲?

8. Bridge problem: two clues

▲ + ▲ = 14

▲ + ■ = 12

Find the value of ■.

9. Challenge: the master balance

▲ + ▲ = ■

■ + ▲ = 15

Find the value of ■.
Hint: Replace ■ *with* (▲ + ▲) *in the second line.*

10. Challenge: reverse thinking

3★ = 24

Now find:

2★ + 5

Chapter Recap

What to remember from this chapter

- The equals sign means both sides have the same value.
- A shape can stand for a number.
- If you know a shape's value, substitute it carefully.
- If you do the same thing to both sides, the balance stays true.
- A quick check helps catch mistakes.

Before moving to Chapter 2

In the next chapter, the shapes will turn into letters such as x, n, and a. The thinking stays the same. Only the notation changes.

Quick Answer Check

Short answers

1. ■ = 9
2. ♡ = 7
3. ● = 11
4. ★ = 5
5. 2 kiwis
6. ■ = 6
7. ▲ = 9
8. ■ = 5
9. ■ = 10
10. 21

Selected fuller solutions

Question 8
From

$$▲ + ▲ = 14$$

we get

$$▲ = 14 \div 2$$
$$▲ = 7.$$

Now substitute into ▲ + ■ = 12:

$$7 + ■ = 12$$
$$■ = 12 - 7$$
$$■ = 5.$$

Question 9
From

$$▲ + ▲ = ■,$$

replace ■ in the second equation:

$$(▲ + ▲) + ▲ = 15.$$

So

$$3▲ = 15$$
$$▲ = 15 \div 3$$
$$▲ = 5.$$

Then

$$\blacksquare = \blacktriangle + \blacktriangle$$

$$\blacksquare = 5 + 5$$

$$\blacksquare = 10.$$

Question 10

From

$$3\bigstar = 24$$

divide both sides by 3:

$$\bigstar = 8.$$

Now substitute into the expression:

$$2\bigstar + 5 = 2 \times 8 + 5 = 16 + 5 = 21.$$

Answer: 21.
Note: the question did not ask for ★ alone. It asked for the value of a whole expression. Always read what is being asked before you start solving.

Algebra with Letters

Variables, Expressions, and Equations

In Chapter 1, shapes stood for numbers. In this chapter, we replace those shapes with letters such as x, n, and a.

The thinking is the same. A letter is just a convenient way to represent a mystery number.

What changes in this chapter?

We now move from pictures and shapes to algebraic language.
That means learning to:

- use letters for unknown numbers,
- tell the difference between an expression and an equation,
- translate words into algebra,
- and avoid common notation traps.

The Main Idea

Concept 1: A letter stands for a number

In algebra, a letter such as x, n, or a stands for a number.
For example:

$$n + 5 = 12$$

This means:

a mystery number plus 5 equals 12

A letter works just like the shapes from Chapter 1. The notation is new, but the idea is the same. In an expression such as $3x$, the number 3 is called the **coefficient**. It tells us how many x's there are.

Two ways a letter is used:

- In an **equation** such as $n + 5 = 12$, the letter stands for one specific mystery number to be found.
- In a **formula** such as $P = 4s$, the letter can take many different values depending on the situation.

In both cases, a letter follows the same rules. The difference is in how the expression is used.

Concept 2: Expression vs equation

These two words are related, but they do **not** mean the same thing.

Term	What it has	Example
Expression	no equals sign	$3n+2$
Equation	an equals sign	$3n+2=14$

An **expression** is just a mathematical phrase.
An **equation** says that two sides have the same value.

Concept 3: Translate words into algebra

To use algebra well, you need to convert words into symbols.

English phrase	**Operation**	**Example**
more than, sum, plus, added to, increased by	$+$	$n+3$
take away, subtract, minus, difference, decreased by	$-$	$n-5$
times, multiplied by, twice, double, triple, product of	$\times$	$2n$
divided by, half of, shared equally, quotient of	$\div$	$\frac{n}{2}$
is, equals, the result is, totals	$=$	$n+3=8$

ThinkLab habit: after translating, read your algebra back in English to check that it still makes sense.

Common trap 1: Invisible multiplication

In algebra, we usually do not write the multiplication sign between a number and a letter.
So:

$$2 \times n$$

is usually written as

$$2n.$$

These mean the same thing.
But be careful:

$$2n \neq 2+n$$

For example, if $n=5$, then

$$2n = 2 \times 5 = 10$$

but

$$2+n = 2+5 = 7.$$

Common trap 2: Less than can flip the order

The phrase *less than* often catches students out.
Look carefully:

Phrase	**Expression**
3 less than x	$x - 3$
a number decreased by 7	$n - 7$
how much less than 12 a number is	$12 - n$

The first row is where most marks are lost. *3 less than x* is $x - 3$, not $3 - x$: you start at x and come down 3. The words arrive in one order and the algebra is written in the other.

Quick check: let $n = 4$.
How much less than 12 is 4?

$$12 - 4 = 8$$

So the correct expression is $12 - n$, not $n - 12$.

Worked Examples

Type A: Translate words into algebra

Example 1: The translator

Problem: Write an expression for each sentence using n for the mystery number.

1. Five more than a number
2. Double a number, then subtract 3
3. How much less than 12 a number is

Solution

1. Five more than a number:

$$n + 5$$

2. Double a number, then subtract 3:

$$2n - 3$$

3. How much less than 12 a number is:

$$12 - n$$

Quick check for part 3
Let $n = 4$:

$$12 - n = 12 - 4$$
$$12 - n = 8$$

That matches the wording.

Type B: Find the value of an expression

Example 2: The value finder

Problem: If $n = 4$, find the value of

$$3n + 2$$

Substitute $n = 4$:

$$\begin{aligned} 3n + 2 &= 3(4) + 2 \\ &= 12 + 2 \\ &= 14 \end{aligned}$$

Answer: the value is **14**.

Note: this is an expression, not an equation to solve. We are finding its value.

Type C: Expression or equation?

Example 3: Sorting the algebra

Problem: Decide whether each one is an expression or an equation.

$$5x + 3 \qquad 2x - 1 = 9 \qquad 4a \qquad n + 7 = 12$$

Solution

$$5x + 3$$

has no equals sign, so it is an **expression**.

$$2x - 1 = 9$$

has an equals sign, so it is an **equation**.

$$4a$$

has no equals sign, so it is an **expression**.

$$n + 7 = 12$$

has an equals sign, so it is an **equation**.

Type D: Write an equation from words

Example 4: A mystery number statement

Problem: "I think of a number, double it, and add 4. The result is 16."
Let the mystery number be n. Write an equation.

Step 1: Double the number

$$2n$$

Step 2: Add 4

$$2n + 4$$

Step 3: The result is 16

$$2n + 4 = 16$$

Answer: The equation is $2n + 4 = 16$.

Training Exercises

1. The translator

Write an expression for each sentence. Use n for the mystery number.

- a) A number plus 8. ____________
- b) Ten less than a number. ____________
- c) Three times a number. ____________
- d) How much less than 10 a number is. ____________

2. Expression or equation?

Write **Ex** for expression or **EQ** for equation next to each one.

- a) $5n + 3$ ________
- b) $2x - 1 = 9$ ________
- c) $4a$ ________
- d) $n + 7 = 12$ ________

3. The evaluator

If $x = 5$, find the value of each expression.

- a) $x + 4$ __________
- b) $2x$ __________
- c) $10 - x$ __________

4. Double trouble

If $y = 6$, find the value of $2y + 3$.

5. The age gap

Ben is b years old.

- a) His sister is 3 years older than Ben. Write an expression for her age.

  ____

- b) Ben's dad is double Ben's age. Write an expression for Ben's dad's age.

  ____

- c) If $b = 10$, how old is Ben's dad? ________

6. Less-than decision drill

Circle the correct expression.

- a) "3 less than x" A) $x - 3$ B) $3 - x$
- b) "x less than 3" A) $x - 3$ B) $3 - x$
- c) "7 less than a number n" A) $n - 7$ B) $7 - n$
- d) "How much less than 7 a number n is" A) $n - 7$ B) $7 - n$

7. The perimeter

A square has side length s.

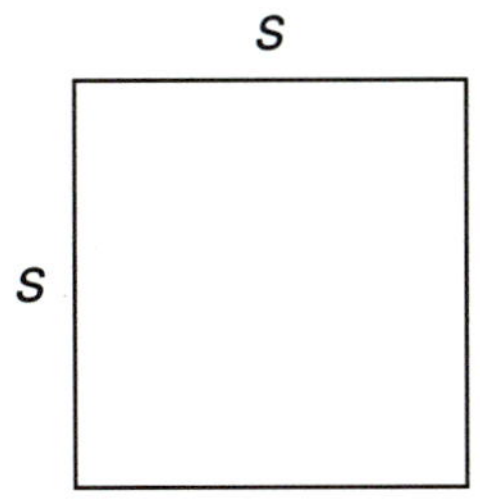

- a) Write an expression for the perimeter. __________
- b) If $s = 4$ cm, what is the perimeter? __________

Working with two variables

The next question has two different letters. Substitute each one separately, just as you would for a single letter.

8. Two variables

If $a = 3$ and $b = 5$, find the value of

$$2a + b$$

Show your working.

9. Write and guess

"I think of a number, triple it, and subtract 7. The result is 23."

- a) Let n be the mystery number. Write an equation.

  Equation: ______________________

- b) Try a few values of n to find the one that works. Show at least two guesses and checks.

My guess for n	Working	Does it work?

In Chapter 4, you will learn a reliable method to solve equations like this without guessing.

Chapter Recap

What to remember from this chapter

- A letter can stand for a number.
- An expression has no equals sign.
- An equation has an equals sign.
- $2n$ means $2 \times n$.
- Be careful with phrases like "less than".
- In an expression like $3n$, the number 3 is called the **coefficient**. It tells us how many n's there are.
- After writing algebra, read it back in English to check it.

Before moving to Chapter 3

In the next chapter, we use bar models to picture algebra before writing equations. This helps turn worded relationships into something you can see and reason about.

Quick Answer Check

Short answers

1. a) $n+8$ b) $n-10$ c) $3n$ d) $10-n$
2. a) Ex b) EQ c) Ex d) EQ
3. a) 9 b) 10 c) 5
4. 15
5. a) $b+3$ b) $2b$ c) 20
6. a) A b) B c) A d) B
7. a) $4s$ b) 16 cm
8. 11
9. a) $3n-7=23$ b) $n=10$

Selected fuller solutions

Question 4
If $y=6$, then

$$\begin{aligned} 2y+3 &= 2(6)+3 \\ &= 12+3 \\ &= 15 \end{aligned}$$

Question 8
If $a=3$ and $b=5$, then

$$\begin{aligned} 2a+b &= 2(3)+5 \\ &= 6+5 \\ &= 11 \end{aligned}$$

Question 9
The words “triple it, and subtract 7” give:

$$3n-7$$

The words “the result is 23” mean:

$$3n-7=23$$

Now try $n=10$:

$$3\times 10-7=30-7=23\checkmark$$

So $n=10$ works.

Bar Models

Seeing the Problem Before Writing the Equation

In Chapter 2, a letter stood for a mystery number. That idea is powerful, but sometimes it helps to *see* what the letter stands for before writing anything down.

A **bar model** is a simple picture made of equal rectangular blocks. Each block represents one unit of a quantity. The picture shows how the quantities in a problem compare, share, or differ. Once the picture is clear, the algebra usually writes itself.

What this chapter is about

Bar models are a bridge between words and algebra. In this chapter, you will learn to:

- draw bar models for sharing, comparison, and ratio problems,
- find the value of one block and use it to answer the question,
- connect a bar model to an equation by letting one block equal x.

The Main Idea

Concept 1: A block is a unit

Every **block** in a bar model represents one equal unit of a quantity. The number of blocks in a row shows how many units that quantity has.

If one person has *twice as many* as another, draw two identical blocks for the first and one block for the second — not one big block and one small block. Blocks must always be drawn the same size, otherwise the picture lies.

A bar model can show three kinds of relationship:

- **Parts of a whole** — one long bar split into pieces.
- **Comparison** — two rows of blocks, where the longer row has extra blocks that hold the difference.
- **Ratio** — two or more rows whose block counts match the ratio.

Concept 2: Find one block

Almost every bar model problem becomes simple once you answer one question:

What is one block worth?

To find it, look for a total or a difference that the problem gives you. Count how many blocks that total or difference covers. Then divide.
Once you know one block, multiply by the number of blocks in any row to find that quantity.

Concept 3: A block is a variable in disguise

A bar model and an algebra equation are the same idea in two different languages.
If you let one block equal x, the bar model turns directly into an equation.
For example, if Tia has 1 block and Sam has 5 blocks, and together they have 120 stamps, then letting one block = x gives:

$$x + 5x = 120$$

$$6x = 120$$

$$x = 20$$

So Tia has 20 stamps and Sam has 100 stamps.
The bar model shows the *structure*. The equation shows the *calculation*. Many strong problem solvers draw the bar first, then write the equation.

Common trap: blocks of different sizes

Every block in a bar model must be drawn the same size. If one person has three times as many marbles as another, draw three identical blocks — not one large block against one small block.
If the blocks are different sizes, the picture no longer matches the problem, and the "find one block" step will give the wrong answer.

Common trap: counting the wrong blocks

Before dividing, always ask: *which blocks does this number belong to?*
A **total** covers every block in every row.
A **difference** covers only the extra blocks in the longer row, not all of them.
For example, if one row is 4 blocks and the other is 7 blocks, then the number 33 means two completely different things:

- if 33 is the **total**, then 11 blocks = 33, so 1 block = 3;
- if 33 is the **difference**, then only 3 blocks = 33, so 1 block = 11.

Mixing these up is the most common bar model mistake.

Worked Examples

Type A: Parts of a total

Example 1: Three-way sharing

Problem: Ella, Finn, and Grace share 180 stickers. Ella has half as many as Finn. Grace has three times as many as Finn. How many stickers does Ella have?

Step 1: Assign blocks.

Let Finn have 2 blocks. Then Ella has 1 block (half of Finn), and Grace has 6 blocks (three times Finn).

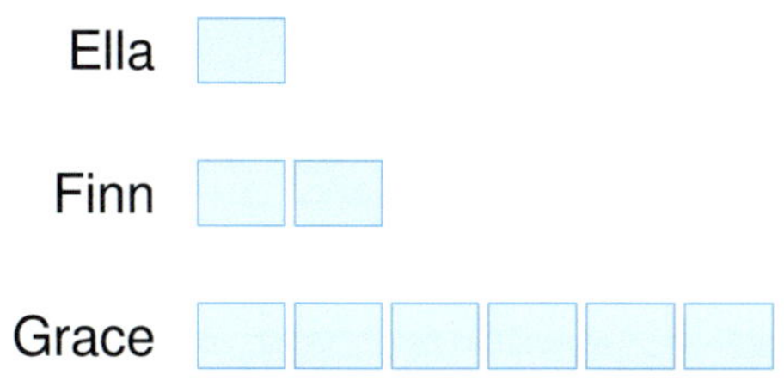

Step 2: Find one block.

Total blocks: $1 + 2 + 6 = 9$.

$$1 \text{ block} = 180 \div 9 = 20$$

Step 3: Answer the question.

Ella has 1 block, so Ella has 20 stickers.

Answer: Ella has 20 stickers.

Type B: Comparison

Example 2: Two ways to solve the same problem

Problem: A tiger weighs 4 times as much as a leopard. The tiger is 105 kg heavier than the leopard. How much does the tiger weigh?

Bar model solution

Leopard: 1 block. Tiger: 4 blocks. The tiger has $4 - 1 = 3$ extra blocks, and those 3 blocks hold the 105 kg difference.

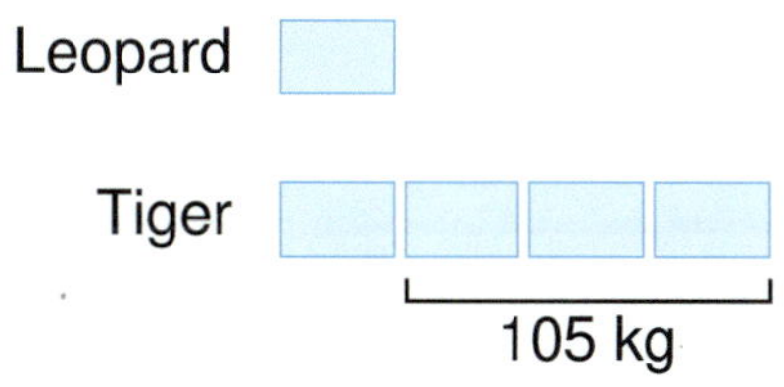

$$3 \text{ blocks} = 105 \text{ kg}, \quad \text{so} \quad 1 \text{ block} = 35 \text{ kg}$$

Tiger = 4 blocks = $4 \times 35 = 140$ kg.

Algebra solution
Let the leopard weigh x kg. Then the tiger weighs $4x$ kg.
The difference is 105 kg:

$$4x - x = 105$$

$$3x = 105$$

$$x = 35$$

Tiger = $4x$ = 140 kg.

Notice that the two solutions use the same three numbers. The bar model shows the *structure*. The algebra shows the *calculation*. One block in the picture is the same as one x in the equation.
Answer: The tiger weighs 140 kg.

Type C: Ratio with a total

Example 3: Smoothie ratio

Problem: A smoothie is made by mixing banana puree, yoghurt, and honey in the ratio 5 : 3 : 2. A large batch uses 360 mL in total. How many millilitres of yoghurt are used?
Step 1: Draw the bars.
Banana: 5 blocks. Yoghurt: 3 blocks. Honey: 2 blocks.
Step 2: Find one block.
Total blocks: $5 + 3 + 2 = 10$.

$$1 \text{ block} = 360 \div 10 = 36 \text{ mL}$$

Step 3: Find yoghurt.
Yoghurt = 3 blocks = 3×36 = 108 mL.
Answer: 108 mL of yoghurt.

Type D: From bar model to equation

Example 4: Let one block equal x

Problem: Ana has three times as many stamps as Ben. Ana has 24 more stamps than Ben. How many stamps does each have?

Step 1: Draw the bars.

Ben: 1 block. Ana: 3 blocks. Ana has 2 extra blocks compared to Ben, and those 2 blocks hold the 24-stamp difference.

Step 2: Turn the bar into an equation.

Let 1 block = x. Then Ben has x stamps and Ana has $3x$ stamps. The difference is 24:

$$3x - x = 24$$

$$2x = 24$$

$$x = 12$$

Step 3: Answer.

Ben has 12 stamps. Ana has $3 \times 12 = 36$ stamps.

Check: $36 - 12 = 24$ ✓

Answer: Ben has 12 stamps, Ana has 36 stamps.

ThinkLab habit: for comparison problems, let one block = x and write a *difference* equation. This works every time the story gives you "how much more" information.

Training Exercises

Draw a bar model for every question. For the later questions, also write the matching equation.

1. **Juice recipe.** A juice mix uses orange and mango in the ratio 3 : 2. If 150 mL of mango is used, how much orange juice is needed?
2. **Stamp collection.** Sam has 5 times as many stamps as Tia. Together they have 120 stamps. How many stamps does Sam have?
3. **Father and daughter.** A father is 4 times as old as his daughter. The father is 36 years older than his daughter. How old is the daughter?
4. **Rectangle.** A rectangle has a perimeter of 56 cm. Its length is three times its width. Find the length and the width.

   Hint: the perimeter contains 2 lengths and 2 widths. How many blocks is that in total?
5. **Chocolate box.** Three siblings share a box of chocolates. Maya takes $\frac{1}{2}$ of the box. Dev takes $\frac{1}{3}$. Sana takes the rest. If Sana receives 10 chocolates, how many chocolates were in the box?
6. **Rope.** A rope is cut into three pieces in the ratio 2 : 5 : 3. The longest piece is 30 cm longer than the shortest piece. What is the total length of the rope?
7. **Container transfer.** Container X holds 4 times as much water as Container Y. After 30 litres are poured from X into Y, both containers hold the same amount. How many litres did Container X hold originally?

Challenge problems

8. **The turtle problem.** The length of a turtle is 60 cm plus one third of its length. How long is the turtle?

   Hint: let the whole turtle be 3 blocks. Then one third of the turtle is 1 block.
9. **Two dogs.** Molly's dog and Holly's dog weigh 80 kg in total. Molly's dog weighs the same as Holly's dog minus a 20 kg bag of dog food. What does Holly's dog weigh?

   Hint: draw two rows. How many extra kg does Holly's dog bar need compared to Molly's?
10. **Equalising transfer.** Lena and Karl have money in the ratio 4 : 7. After Lena earns an extra $24 and Karl spends $12, they have the same amount. How much did Karl have to start with?

Chapter Recap

What to remember from this chapter

- Every block in a bar model is the same size and stands for one unit.
- The core question is always: *what is one block worth?*
- A total covers every block in every row. A difference covers only the extra blocks.
- A bar model and an algebra equation are the same idea in two languages. Let one block = x, and the picture becomes an equation.

Before moving to Chapter 4

So far, you have met the language of algebra (Chapter 2) and the picture that goes with it (Chapter 3). In the next chapter, we learn how to *solve* equations systematically by undoing operations in the right order. Bar models will often still be useful — draw first, solve second.

Quick Answer Check

Short answers

1. 225 mL
2. 100 stamps
3. 12 years old
4. Width 7 cm, length 21 cm
5. 60 chocolates
6. 100 cm
7. 80 litres
8. 90 cm
9. 50 kg
10. $84

Selected fuller solutions

Question 4 (Rectangle).
Let the width be 1 block and the length be 3 blocks. Drawing the rectangle with those blocks:

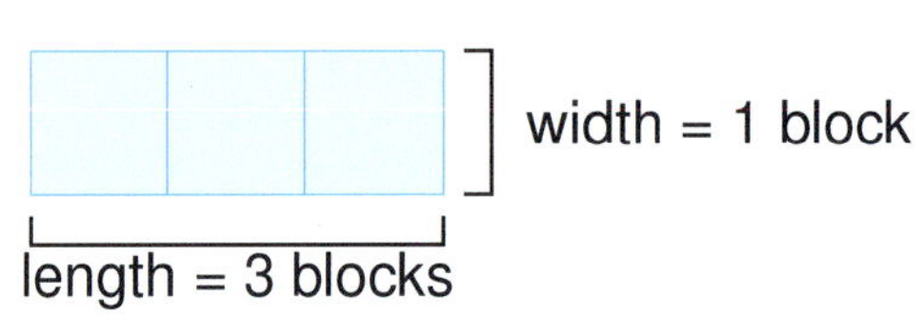

The perimeter is 2 lengths + 2 widths, which is $2 \times 3 + 2 \times 1 = 8$ blocks in total. That perimeter is 56 cm, so

$$8 \text{ blocks} = 56 \text{ cm}, \text{ so } 1 \text{ block} = 7 \text{ cm}.$$

Width = 7 cm, length = $3 \times 7 = 21$ cm.

Check: $2(7) + 2(21) = 14 + 42 = 56$ cm. ✓
Answer: Width = 7 cm, length = 21 cm.

Question 5 (Chocolate box).
Draw the whole box as one bar, split into the three shares.

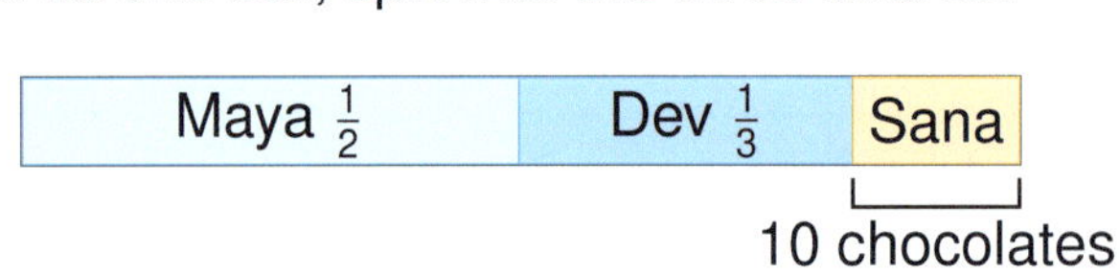

The bar is the whole box. Maya's slice is $\frac{1}{2}$ and Dev's is $\frac{1}{3}$, so Sana's slice is whatever is left:

$$1 - \frac{1}{2} - \frac{1}{3} = \frac{6}{6} - \frac{3}{6} - \frac{2}{6} = \frac{1}{6}.$$

So Sana's slice is $\frac{1}{6}$ of the box, and that slice is 10 chocolates. The whole box must be 6 of those slices:

$$6 \times 10 = 60 \text{ chocolates.}$$

Answer: 60 chocolates.

Question 7 (Container transfer).
Choosing the block size. If we set Y = 1 block and X = 4 blocks, the total is 5 blocks. After the transfer each container holds half of that, which is $2\frac{1}{2}$ blocks — awkward to work with.
We can avoid the fraction by doubling: set Y = 2 blocks and X = 8 blocks. The ratio is still 4 to 1, and now the total is 10 blocks, which splits evenly.

Together they hold 10 blocks, and that total does not change when water is poured from one container to the other. After the transfer both containers are equal, so each holds 5 blocks.
X went from 8 blocks down to 5 blocks, a drop of 3 blocks. That drop equals 30 litres, so

$$3 \text{ blocks} = 30 \text{ L}, \text{ so } 1 \text{ block} = 10 \text{ L}.$$

X originally held $8 \times 10 = 80$ litres.

ThinkLab habit: when the bar has to split evenly later, pick a block size that makes the split whole. Doubling (or tripling) the blocks never changes the ratio — it only changes the unit.

Question 8 (Turtle).
Following the hint, draw the whole turtle as 3 blocks. The problem says the turtle's length is 60 cm plus one third of itself — so the same turtle can also be drawn as 1 block plus a 60 cm piece.

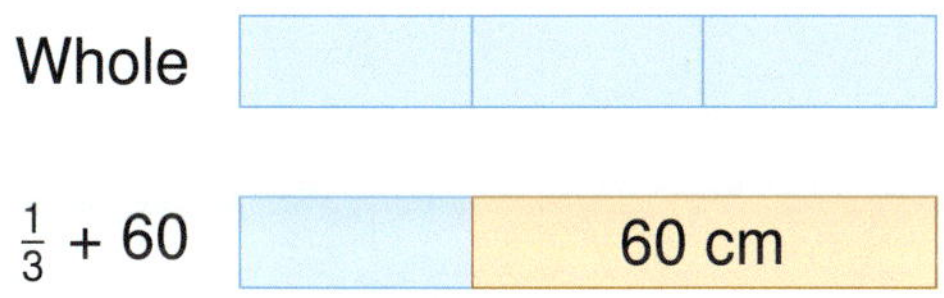

Both rows show the same length (the whole turtle), so the 60 cm piece must equal the 2 extra blocks on top. So

$$2 \text{ blocks} = 60 \text{ cm}, \text{ so } 1 \text{ block} = 30 \text{ cm}.$$

The whole turtle is $3 \times 30 = 90$ cm.

As an equation. Let the length be x. Then

$$x = 60 + \tfrac{1}{3}x, \text{ so } \tfrac{2}{3}x = 60, \text{ giving } x = 90 \text{ cm}.$$

Answer: The turtle is 90 cm long.

Question 9 (Two dogs).
"Molly's dog weighs the same as Holly's dog minus 20 kg" means Holly's dog is 20 kg heavier than Molly's dog. Draw Molly as 1 block and Holly as 1 block plus an extra 20 kg piece.

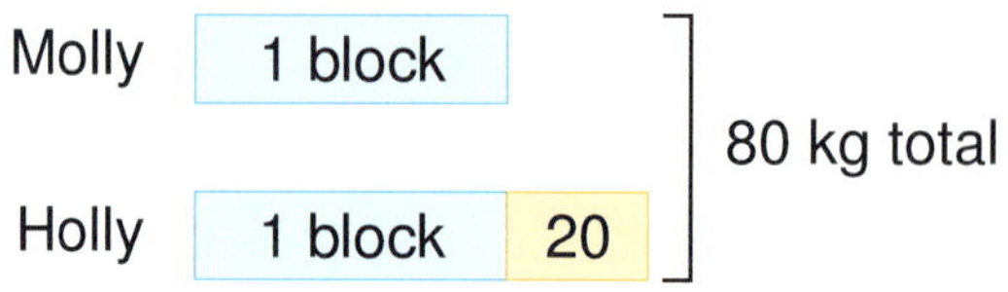

The two bars together hold 2 blocks + 20 kg, and that total is 80 kg. So

$$2 \text{ blocks} + 20 = 80, \text{ so } 2 \text{ blocks} = 60, \text{ so } 1 \text{ block} = 30 \text{ kg}.$$

Holly's dog is 1 block + 20 kg, which is 30 + 20 = 50 kg.
Answer: Holly's dog weighs 50 kg.

Question 10 (Equalising transfer).
Lena and Karl start in the ratio 4 : 7. Draw Lena as 4 blocks and Karl as 7 blocks.

Now apply the changes. Lena earns \$24, so her total becomes 4 blocks + 24. Karl spends \$12, so his total becomes 7 blocks − 12.

After the changes the two amounts are equal, so

$$4 \text{ blocks} + 24 = 7 \text{ blocks} - 12.$$

Move the blocks to one side and the numbers to the other:

$$24 + 12 = 7 \text{ blocks} - 4 \text{ blocks}, \text{ so } 3 \text{ blocks} = 36.$$

That gives 1 block = 12 dollars. Karl started with $7 \times 12 = 84$ dollars.

As an equation. Let 1 block = x. Then Lena had $4x$ and Karl had $7x$. After the changes:

$$4x + 24 = 7x - 12, \text{ so } 3x = 36, \text{ giving } x = 12.$$

Karl started with $7x = 84$ dollars.
Answer: Karl had \$84 to start with.

Solving Equations

Undoing Step by Step

In Chapters 2 and 3, you learned how to write algebra from words and picture relationships with bar models. In this chapter, the focus shifts to solving equations.

The goal is simple: we want to get the mystery number on its own. To do that, we use **inverse operations**, which means undoing what has been done to the variable.

What this chapter is about

When you solve an equation, you are trying to isolate the variable.
Useful habits in this chapter:

- keep the equation balanced,
- undo operations in the reverse order,
- show one clear step at a time,
- and always check your answer at the end.

The Main Idea

Concept 1: Inverse operations are opposites

To solve for a variable, we undo what is happening to it.

If you see...	Undo by...	Example
$+5$	subtract 5	$x + 5 = 10$
-3	add 3	$x - 3 = 7$
$\times 2$ or $2x$	divide by 2	$2x = 12$
$\div 4$ or $\frac{x}{4}$	multiply by 4	$\frac{x}{4} = 3$

The key idea is: do the opposite operation to both sides.

Concept 2: Reverse the steps

Suppose an expression is built in this order:

- start with x,
- multiply by 3,
- then add 5.

That gives the expression $3x + 5$. Now suppose we're told this expression equals some number — say, $3x + 5 = 17$. To solve it, we undo the steps in reverse order:

- subtract 5 first,
- then divide by 3.

ThinkLab habit: undo in the reverse order that the operations happened.

Why does order matter? A quick demonstration.
Suppose someone tries to solve $3x + 5 = 17$ by dividing by 3 first:

$$x + \frac{5}{3} = \frac{17}{3}$$

This is technically valid, but now the numbers are messy fractions. Starting with subtraction is much cleaner:

$$3x = 12 \qquad \text{then} \qquad x = 4.$$

Common trap: dividing too early

Some students see

$$2x + 4 = 10$$

and divide everything by 2 straight away in their head, but then write

$$x + 4 = 5.$$

That is wrong.
If you divide both sides by 2 correctly, then

$$\frac{2x + 4}{2} = \frac{10}{2}$$

$$x + 2 = 5.$$

A safer method here is to undo the addition first:

$$2x + 4 = 10$$

$$2x = 6$$

$$x = 3.$$

Worked Examples

Type A: One-step equations

Example 1: Simple undoing

Problem: Solve for *n*.

$$n - 8 = 12$$

Undo the subtraction by adding 8 to both sides:

$$n = 12 + 8$$

$$n = 20$$

Quick check

$$20 - 8 = 12$$

Answer: $n = 20$.

Type B: Multiply then add

Example 2: Two-step equation

Problem: Solve for *x*.

$$3x + 5 = 17$$

Step 1: Undo the addition

$$3x = 17 - 5$$

$$3x = 12$$

Step 2: Undo the multiplication

$$x = 12 \div 3$$

$$x = 4$$

Quick check

$$3 \times 4 + 5 = 12 + 5$$

$$3 \times 4 + 5 = 17$$

Answer: $x = 4$.

Type C: Fractions

Example 3: Fraction solver

Problem: Solve for *x*.

$$\frac{x}{3} + 2 = 9$$

Step 1: Undo the addition

$$\frac{x}{3} = 9 - 2$$

$$\frac{x}{3} = 7$$

Step 2: Undo the division by 3

$$x = 7 \times 3$$

$$x = 21$$

Quick check

$$\frac{21}{3} + 2 = 7 + 2$$

$$\frac{21}{3} + 2 = 9$$

Answer: $x = 21$.

Type D: Write and solve from words

Example 4: A word problem

Problem: "A student thinks of a number, multiplies it by 2, and then adds 2. The answer is 20."

Step 1: Write the equation

$$2n + 2 = 20$$

Step 2: Undo the addition

$$2n = 20 - 2$$

$$2n = 18$$

Step 3: Undo the multiplication

$$n = 18 \div 2$$

$$n = 9$$

Answer: $n = 9$.

Training Exercises

1. One-step warm-up

Solve:

- a) $x + 7 = 15$ $x =$ ______
- b) $x - 9 = 4$ $x =$ ______
- c) $3x = 21$ $x =$ ______
- d) $\frac{x}{2} = 8$ $x =$ ______

2. Two-step training

Solve:

- a) $2x + 3 = 13$ $x =$ ______
- b) $3n - 5 = 10$ $n =$ ______
- c) $4y - 3 = 21$ $y =$ ______

3. Fraction training

Solve:

- a) $\frac{x}{2} + 3 = 8$
- b) $\frac{n}{3} - 1 = 4$

4. Ages

Ben's dad is 3 times as old as Ben. Ben's dad is 36 years old.

- Write an equation using b for Ben's age.
- Solve for b.

5. Shopping

Ava bought 3 giant cookies and a $2 drink. The total cost was $11.

- Write an equation using c for the cost of one cookie.
- Solve for c.

6. Consecutive numbers

The sum of three consecutive numbers is 33. Find the numbers.

Method A:

$$n,\ n+1,\ n+2$$

Method B:

$$n-1,\ n,\ n+1$$

Use either method to:

- a) write an equation,
- b) solve it,
- c) write the three numbers.

Notice: the two methods use different meanings for n, but they still lead to the same three numbers.

7. Harder equation

Solve:

$$\frac{x}{4}+5=11$$

8. Extension: variables on both sides

So far, every equation had the variable on one side only. In those cases, the inverse-operations method works cleanly. When the variable appears on **both sides**, use the balance method: move all variable terms to one side first, then solve.

Solve:

$$4x-7=2x+9$$

Hint: subtract $2x$ *from both sides first. This collects all the* x*-terms on the left.*

Chapter Recap

What to remember from this chapter

- The goal is to get the variable on its own.
- Use inverse operations to undo what is happening.
- Undo the steps in reverse order.
- Keep both sides balanced.
- For word problems, define the variable first, then write the equation before solving.
- Check your answer by substituting it back.

Before moving to Chapter 5

So far, the main goal has been solving equations. In the next chapter, we also learn how to simplify algebraic expressions by collecting like terms and working with brackets.

Quick Answer Check

Short answers

1. a) 8 b) 13 c) 7 d) 16
2. a) 5 b) 5 c) 6
3. a) 10 b) 15
4. $3b = 36$, so $b = 12$
5. $3c + 2 = 11$, so $c = 3$
6. 10, 11, 12 *(Method A gives $n = 10$; Method B gives $n = 11$ — different n, same three numbers)*
7. 24
8. 8

Selected fuller solutions

Question 5

Let c be the cost of one cookie.
Three cookies and a \$2 drink cost \$11, so:

$$3c + 2 = 11$$

Subtract 2 from both sides:

$$3c = 11 - 2$$

$$3c = 9$$

Divide both sides by 3:

$$c = 9 \div 3$$

$$c = 3$$

So one cookie costs \$3.

Quick check:

$$3 \times 3 + 2 = 9 + 2 = 11\checkmark$$

Question 8

$$4x - 7 = 2x + 9$$

Subtract $2x$ from both sides:

$$2x - 7 = 9$$

Add 7 to both sides:

$$2x = 16$$

Divide both sides by 2:

$$x = 8$$

Simplifying Expressions

Like Terms, Brackets, and Expansion

In the previous chapters, the main goal was usually to find the value of a mystery number. In this chapter, the goal is often different: we want to rewrite an expression in a simpler form.

That means learning to:

- combine like terms,
- expand brackets carefully,
- handle negative brackets,
- and, as an optional extension, recognise when a common factor can be taken out.

A new kind of algebra task

Not every algebra question asks you to solve an equation.
Sometimes you are asked to **simplify** an expression. That means rewriting it in a cleaner form without changing its value.

Part A: Collecting Like Terms

Concept 1: Like terms must be the same type

You can only add or subtract terms if they are the same type.
For example:

$$2a + 3a = 5a$$

but

$$2a + 3b$$

cannot be combined, because a and b are different variables.

ThinkLab rule: like terms must have the same letter and the same power.
So:

$$x \text{ and } x$$

are like terms, but

$$x \text{ and } x^2$$

are not.

Common trap: the sign belongs to the term

When simplifying, the plus or minus sign travels with the term.
For example:

$$8x - 2y - 3x$$

Some students spot the two x terms and add them, leaving the minus sign behind:

$$8x + 3x - 2y = 11x - 2y.$$

That is wrong. The third term is $-3x$, not $3x$, so the minus sign has to travel with it.
Group the like terms carefully:

$$(8x - 3x) - 2y$$

$$5x - 2y$$

ThinkLab habit: read the expression with all plus signs first, $8x + (-2y) + (-3x)$.
Then the terms can be moved into any order and no sign is ever lost.

Worked Examples: Like Terms

Example 1: Tidy up

Problem: Simplify

$$4a + 5b + 2a - 3b$$

Group like terms:

$$(4a + 2a) + (5b - 3b)$$

Simplify each group:

$$6a + 2b$$

Answer: $6a + 2b$.

Example 2: Numbers join in too

Problem: Simplify

$$3 + 2a + 5 - a$$

Group like terms. The number terms (constants) are like terms with each other, and the a terms are like terms with each other:

$$(2a - a) + (3 + 5)$$

Simplify each group:

$$a + 8$$

Answer: $a + 8$.

Example 3: Different powers

Note: x^2 means $x \times x$. It is a different type of term from x, so they cannot be combined.

Problem: Simplify

$$2x + 3x^2 + 5x - x^2$$

Group like terms:

$$(2x + 5x) + (3x^2 - x^2)$$

Simplify:

$$7x + 2x^2$$

It is conventional to write higher powers first, so:

$$2x^2 + 7x$$

Answer: $2x^2 + 7x$.

Part B: Brackets and Distribution

Concept 2: Distributive law

A number outside brackets must multiply every term inside the brackets.
For example:

$$2(x + 3)$$

means:

$$2 \times x + 2 \times 3$$

so:

$$2(x + 3) = 2x + 6.$$

ThinkLab habit: multiply the outside number by each term inside, one term at a time.

Common trap: a minus in front of brackets

A leading minus sign means multiply the whole bracket by -1.
For example:

$$-(x - 5)$$

means:

$$(-1)(x - 5)$$

so:

$$-(x - 5) = -x + 5.$$

The signs inside the bracket flip.
A very common slip is to flip only the first sign and write

$$-x - 5.$$

That is wrong: the -1 multiplies **every** term inside, and $(-1) \times (-5) = +5$.

Worked Examples: Brackets

Example 4: Expand a bracket

Problem: Expand

$$3(2x+5)$$

Multiply 3 by each term:

$$3 \times 2x + 3 \times 5$$

Simplify:

$$6x + 15$$

Answer: $6x + 15$.

Example 5: Expand and simplify

Problem: Simplify

$$3(x+2) + 2(x-1)$$

Expand the first bracket:

$$3(x+2) = 3x + 6$$

Expand the second bracket:

$$2(x-1) = 2x - 2$$

Now combine:

$$3x + 6 + 2x - 2$$

Collect like terms:

$$(3x + 2x) + (6 - 2)$$

Simplify:

$$5x + 4$$

Answer: $5x + 4$.

Example 6: Negative brackets

Problem: Simplify

$$-2(x+3) + (x-4)$$

Expand the first bracket:

$$-2(x+3) = -2x - 6$$

Keep the second bracket:

$$(x-4) = x - 4$$

Now combine:

$$-2x - 6 + x - 4$$

Collect like terms:

$$(-2x + x) + (-6 - 4)$$

Simplify:

$$-x - 10$$

Answer: $-x - 10$.

Part C: Reverse Distribution (Optional Extension)

Concept 3: Factoring out a common factor

Sometimes an expression can be rewritten more neatly by taking out a common factor.
For example:

$$8x + 12$$

Both terms are divisible by 4, so:

$$8x + 12 = 4(2x + 3)$$

This is called **factoring**.
You do not always need this in upper primary algebra, but it is a useful extension skill.

Training Exercises

1. Like terms

Simplify:

- a) $3x + 2x + 5x$ ____________
- b) $10y - 2y + y$ ____________
- c) $5a + 2b + 3a$ ____________

2. Signs matter

Simplify:

- a) $9x - 4y - 2x$ ____________
- b) $5 + 2a + 3 - a$ ____________
- c) $2x + 5 - 2x$ ____________

3. Different powers

Simplify:

$$4x^2 + 3x - x^2 + 6x$$

4. Expand

Expand:

- a) $2(x + 4)$ ______
- b) $3(2n - 5)$ ______
- c) $5(10 - x)$ ______

5. Negative brackets

Expand:

- a) $-2(x + 3)$ ______
- b) $-(x - 5)$ ______

6. Expand and collect

Simplify:

$$4(x - 2) + 3(x + 5)$$

7. Subtracting a bracket

Simplify:

$$2(3x - 4) - 3(x + 2)$$

Hint: expand both brackets first, then collect like terms.

8. Optional extension: factoring

Factor completely:

- a) $6x + 9$
- b) $12a - 18$

Chapter Recap

What to remember from this chapter

- Like terms must have the same letter and the same power.
- Keep the sign with the term.
- To expand a bracket, multiply the outside number by every term inside.
- A minus in front of brackets flips the signs inside.
- Expand first, then collect like terms.

Before moving to Chapter 6

So far, you have worked with equations and algebraic expressions separately. In the next chapter, we bring these ideas together in word problems involving totals, comparisons, and consecutive numbers.

Quick Answer Check

Short answers

1. a) $10x$ b) $9y$ c) $8a + 2b$
2. a) $7x - 4y$ b) $a + 8$ c) 5
3. $3x^2 + 9x$
4. a) $2x + 8$ b) $6n - 15$ c) $50 - 5x$
5. a) $-2x - 6$ b) $-x + 5$
6. $7x + 7$
7. $3x - 14$
8. a) $3(2x + 3)$ b) $6(2a - 3)$

Selected fuller solutions

Question 3

$$4x^2 + 3x - x^2 + 6x$$

Group like terms:

$$(4x^2 - x^2) + (3x + 6x)$$

Simplify:

$$3x^2 + 9x$$

Question 6

$$4(x - 2) + 3(x + 5)$$

Expand the first bracket:

$$4(x - 2) = 4x - 8$$

Expand the second bracket:

$$3(x + 5) = 3x + 15$$

Combine:

$$4x - 8 + 3x + 15$$

Collect like terms:

$$(4x + 3x) + (-8 + 15)$$

Simplify:

$$7x + 7$$

Question 7

$$2(3x - 4) - 3(x + 2)$$

Expand the first bracket:

$$2(3x - 4) = 6x - 8$$

Expand the second bracket carefully:

$$-3(x + 2) = -3x - 6$$

Now combine:

$$6x - 8 - 3x - 6$$

Collect like terms:

$$(6x - 3x) + (-8 - 6)$$

Simplify:

$$3x - 14$$

Question 8

a) The largest number that divides both 6 and 9 is 3:

$$6x + 9 = 3(2x + 3)$$

b) The largest number that divides both 12 and 18 is 6:

$$12a - 18 = 6(2a - 3)$$

The word *completely* is doing real work here. Taking out a smaller factor is not wrong, but it is not finished:

$$12a - 18 = 2(6a - 9)$$

There is still a 3 hiding inside the bracket, because 6 and 9 share a factor of 3.

Check: expand your answer and see whether you get back to the start.

$$6(2a - 3) = 12a - 18 \checkmark$$

Word Problems with Algebra

Totals, Comparisons, and Consecutive Numbers

In the earlier chapters, you learned how to write algebra, draw bar models, solve equations, and simplify expressions. In this chapter, those ideas come together in word problems.

The key challenge is no longer just doing the algebra. The harder part is deciding:

- what the variable should represent,
- how the quantities are related,
- and which equation matches the story.

A useful strategy for word problems

When a word problem feels busy, slow down and ask:

- What is the unknown?
- What letter will I use for it?
- What do the other quantities become?
- What equation connects them?

Before writing any algebra, try drawing a bar model (Chapter 3). Many word problems become easier to read once the picture is in front of you.
A good algebra solution usually starts with a clear variable definition.

The Main Idea

Concept 1: Choose a good variable

In many problems, one quantity is simpler than the others. That is often the best one to choose as the variable.
For example, if Tom has 10 more marbles than Sarah, it is natural to let

$$\text{Sarah} = x$$

so that

$$\text{Tom} = x + 10.$$

If one quantity is twice or three times another, it is often best to choose the smaller one as the variable.

Concept 2: Build the equation from the relationships

Once the variable is chosen, express the other quantities in terms of it.
Then use the information in the problem to write one equation.
For example, if two quantities add to 50:

$$x + (x + 10) = 50.$$

The algebra becomes much easier once the structure is written clearly.

Concept 3: Consecutive numbers

Consecutive numbers are numbers that come one after another.
Three consecutive numbers can be written as:

$$n,\ n+1,\ n+2$$

Sometimes a more convenient form is:

$$n-1,\ n,\ n+1$$

This second form is often useful because the middle number is easy to identify.

Common trap: choosing an awkward variable

A problem can still be solved with a complicated variable choice, but it becomes harder than necessary.
For example, if Tom has 10 more than Sarah, it is much cleaner to let Sarah be x and Tom be $x + 10$ than to choose a more awkward expression first.
If you let Tom be x instead, then Sarah is $x - 10$, and every line after that carries a subtraction. The answer still comes out, but there are more places to slip.
A good default is to let x be the **smaller** quantity, so the other one is built by adding rather than subtracting.
ThinkLab habit: choose the variable that makes the rest of the problem easy to describe.

Worked Examples

Type A: Comparison plus total

Example 1: Two people, one total

Problem: Tom and Sarah have \$50 altogether. Tom has \$10 more than Sarah. How much money does each person have?
Step 1: Choose a variable
Let Sarah have x dollars.
Then Tom has $x + 10$ dollars.
Step 2: Write the equation

Together they have \$50, so:

$$x + (x + 10) = 50$$

Step 3: Simplify

$$2x + 10 = 50$$

Step 4: Solve

$$2x = 50 - 10$$

$$2x = 40$$

$$x = 20$$

Step 5: Interpret the answer
Sarah has \$20.
Tom has 20 + 10 = 30 dollars.
Answer: Sarah has \$20 and Tom has \$30.

Bar model view

Draw two rows. Sarah has 1 block. Tom has 1 block plus a small strip of \$10. Together the two rows hold 2 blocks +\$10 = \$50, so 2 blocks = \$40 and 1 block = \$20.
That matches the algebra exactly: "1 block" in the picture is the same as "x" in the equation. Sarah has \$20 and Tom has \$30.

Type B: Total with expressions

Example 2: Three friends and marbles

Problem: Three friends have 45 marbles in total.

- Friend A has x,
- Friend B has $x + 5$,
- Friend C has $2x$.

Find x.
Step 1: Write the total equation

$$x + (x + 5) + 2x = 45$$

Step 2: Simplify

$$4x + 5 = 45$$

Step 3: Solve

$$4x = 45 - 5$$

$$4x = 40$$

$$x = 10$$

Interpret the answer:
Friend A has $x = 10$ marbles.

Friend B has $x + 5 = 10 + 5 = 15$ marbles.
Friend C has $2x = 2 \times 10 = 20$ marbles.
Quick check: $10 + 15 + 20 = 45$ ✓
Answer: Friend A: 10, Friend B: 15, Friend C: 20.

Type C: Consecutive numbers

Example 3: Three consecutive numbers

Problem: The sum of three consecutive numbers is 33. Find the numbers.
Method 1: Standard form
Let the numbers be n, $n + 1$, and $n + 2$.
Write the equation:

$$n + (n + 1) + (n + 2) = 33$$

Simplify:

$$3n + 3 = 33$$

Solve:

$$3n = 33 - 3$$

$$3n = 30$$

$$n = 10$$

So the numbers are:

$$10,\ 11,\ 12$$

Method 2: Middle number form
Let the numbers be $n - 1$, n, and $n + 1$.
Then the sum is:

$$(n - 1) + n + (n + 1) = 33$$

Simplify:

$$3n = 33$$

Solve:

$$n = 11$$

So the numbers are:

$$10,\ 11,\ 12$$

Answer: The three numbers are 10, 11, 12.

Bar model view

Draw three equal blocks. The second block has a small +1 strip attached; the third has a small +2 strip.
The total of all three rows is 33. The three equal blocks account for $33 - 1 - 2 = 30$, so each block is 10. The three numbers are 10, 11, 12.
This is also why the "middle number" method is so clean. If you shift the blocks to $n - 1$, n, and $n + 1$, the small strips on the two sides cancel out. Three equal blocks

account for the whole total, and you get n directly.

Type D: Perimeter problem

Example 4: Rectangle perimeter

Problem: A rectangular swimming pool has perimeter 40 m. The length is 4 m longer than the width. Find the length.
Step 1: Define the variable
Let the width be w.
Then the length is $w + 4$.
Step 2: Write the perimeter equation
Perimeter of a rectangle:

$$2(\text{width} + \text{length})$$

So:

$$2(w + (w + 4)) = 40$$

Step 3: Simplify

$$2(2w + 4) = 40$$

$$4w + 8 = 40$$

Step 4: Solve

$$4w = 40 - 8$$

$$4w = 32$$

$$w = 8$$

Step 5: Find the length

$$w + 4 = 8 + 4$$

$$w + 4 = 12$$

Answer: The length is **12** m.

Type E: Fraction-based word problem

Example 5: Marbles left over

Problem: A jar is filled with some marbles. Sam takes $\frac{1}{5}$ of them. Then Lina takes 12 marbles. There are 48 marbles left. How many marbles were there at the start?
Step 1: Define the variable
Let the starting number of marbles be x.
Step 2: After Sam takes $\frac{1}{5}$, what remains?
If $\frac{1}{5}$ is taken away, then $\frac{4}{5}$ remains:

$$\frac{4}{5}x$$

Step 3: Lina then takes 12 more

So the remaining amount is:

$$\frac{4}{5}x - 12$$

This equals 48:

$$\frac{4}{5}x - 12 = 48$$

Step 4: Solve

$$\frac{4}{5}x = 48 + 12$$

$$\frac{4}{5}x = 60$$

We need to undo "multiply by $\frac{4}{5}$". Do this in two steps: first multiply both sides by 5, then divide by 4.
Multiply both sides by 5:

$$4x = 60 \times 5$$

$$4x = 300$$

Divide both sides by 4:

$$x = 300 \div 4$$

$$x = 75$$

Answer: There were **75** marbles at the start.

Bar model view

Draw the starting marbles as 5 equal blocks. Sam takes 1 block, Lina then takes 12 marbles from what remains, and 48 marbles are left.

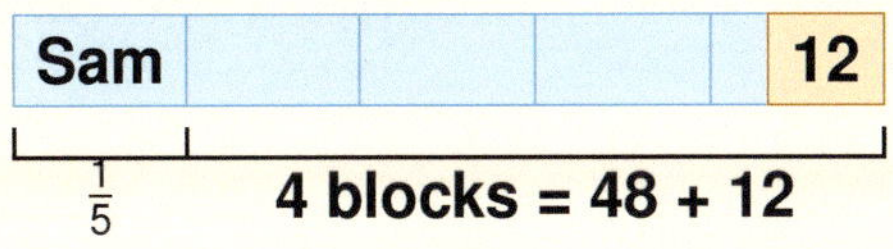

The 4 blocks after Sam contain the 48 marbles left plus the 12 Lina took, so

$$4 \text{ blocks} = 48 + 12 = 60, \text{ so } 1 \text{ block} = 15 \text{ marbles.}$$

The starting amount was $5 \times 15 = 75$ marbles.

The bar model skips the fraction calculation entirely. The "multiply both sides by 5, then divide by 4" step in the algebra solution is hidden inside the picture — the blocks already *are* fifths of the whole.

Training Exercises

1. Pick the pilot

Write expressions only. Do not solve.
Ali is 5 years older than Ben. Cara is twice as old as Ben.
Let Ben = x.

- Ali = ________________
- Cara = ________________

2. Sticker collection

Three friends collect stickers.

- Friend A has x stickers.
- Friend B has twice as many as Friend A.
- Friend C has $x + 8$ stickers.

Together they have 60 stickers.

- Write an equation.
- Solve for x.
- State how many stickers each friend has.

3. Consecutive numbers

The sum of three consecutive numbers is 45.
Let the numbers be n, $n + 1$, and $n + 2$.
Find the largest number.

4. Library shelves

A school library has two shelves of books. The top shelf has 3 times as many books as the bottom shelf. Together, the two shelves hold 84 books.

- Let the number of books on the bottom shelf be x. Write an expression for the number on the top shelf.
- Write an equation and solve it.
- State how many books are on each shelf.

5. Ages

A father is 5 times as old as his son. The father is 32 years older than the son.
How old is the son?

6. Bookshop sales

A bookshop sells 15 more books on Saturday than double its usual weekday sales. This is the same as five times its usual weekday sales minus 6 books.
How many books does the bookshop usually sell on a weekday?
Strategy: let weekday sales = x*. Write two different expressions for Saturday sales, then set them equal to each other.*

7. Packages

Two packages weigh 44 kg altogether. The heavier package weighs 8 kg more than the lighter one.
How much does the heavier package weigh?

8. Basketball score

Alex scores four times as many points as in his previous game. Later, the official score is corrected by subtracting 13 points. Alex's final score is then 8 more than his previous game's score.
How many points did Alex score in his previous game?

9. Library books

A library lends out one quarter of its books, then lends out 7 more books. After that, 14 books remain on the shelves.
How many books does the library have?

Chapter Recap

What to remember from this chapter

- Start by defining the variable clearly.
- Express the other quantities in terms of that variable.
- Use the words in the problem to build one equation.
- Solve carefully, one step at a time.
- Check that your answer makes sense in the original story.

Before moving to Chapter 7

So far, algebra has helped us describe and solve unknown quantities. In the next chapter, we use algebra to describe growing patterns and sequences, and to jump straight to faraway terms.

Quick Answer Check

Short answers

1. Ali = $x + 5$, Cara = $2x$
2. $x = 13$; Friend A: 13, Friend B: 26, Friend C: 21
3. 16
4. Bottom shelf: 21 books; Top shelf: 63 books
5. 8 years old
6. 7 books
7. 26 kg
8. 7 points
9. 28 books

Selected fuller solutions

Question 4
Let the bottom shelf have x books. Then the top shelf has $3x$ books.
Total equation:

$$x + 3x = 84$$
$$4x = 84$$
$$x = 21$$

So the bottom shelf has 21 books and the top shelf has $3 \times 21 = 63$ books.
Quick check: $21 + 63 = 84$ ✓

Question 5
Let the son's age be x.
Then the father's age is $5x$.
The father is 32 years older, so:

$$5x = x + 32$$

Subtract x from both sides:

$$4x = 32$$

Divide by 4:

$$x = 8$$

So the son is **8** years old.

Question 7
Let the lighter package weigh x kg.
Then the heavier package weighs $x + 8$ kg.
Together they weigh 44 kg:

$$x + (x + 8) = 44$$

Simplify:

$$2x + 8 = 44$$

Solve:

$$2x = 44 - 8$$

$$2x = 36$$

$$x = 18$$

Therefore, the heavier package weighs $18 + 8 =$ **26** kg.

Question 8
Let the previous game's score be x.
The new score before the penalty is $4x$.
After 13 points are taken away, the final score is $4x - 13$.
This equals the previous score plus 8, which is $x + 8$.
So:

$$4x - 13 = x + 8$$

Subtract x from both sides:

$$3x - 13 = 8$$

Add 13 to both sides:

$$3x = 21$$

Divide by 3:

$$x = 7$$

So the previous game's score was **7** points.

Question 9
Let the starting number of books be x.
After lending out one quarter, the remaining books are $\frac{3}{4}x$.
After 7 more books are lent out, the remaining books are $\frac{3}{4}x - 7$.
This equals 14:

$$\frac{3}{4}x - 7 = 14$$

Add 7 to both sides:

$$\frac{3}{4}x = 21$$

Multiply both sides by 4:

$$3x = 84$$

Divide both sides by 3:

$$x = 28$$

Quick check: $\frac{3}{4} \times 28 - 7 = 21 - 7 = 14$ ✓
Answer: The library has 28 books.

Patterns and Sequences

Finding Rules and the nth Term

In the earlier chapters, algebra helped us describe unknown numbers and solve equations. In this chapter, algebra helps us describe **patterns**.

The big idea is this: instead of writing out every term one by one, we want to find a rule that works for **any** term.

What this chapter is about

When a sequence follows a clear pattern, algebra helps us:

- describe how the pattern grows,
- find faraway terms quickly,
- write a general rule,
- and test whether a number belongs to the sequence.

The Main Idea

Concept 1: Look for the starting number and the step size

A sequence is an ordered list of numbers.
For example:

$$4,\ 7,\ 10,\ 13,\ 16, \ldots$$

This sequence:

- starts at 4,
- goes up by 3 each time.

That constant change of +3 is called the **common difference**.

Concept 2: The nth term rule

For an arithmetic sequence, the nth term can be found using:

$$\text{nth term} = \text{first term} + (n - 1) \times \text{difference}$$

So if a sequence starts at 4 and increases by 3 each time, then:

$$\text{nth term} = 4 + (n - 1) \times 3$$

This can also be simplified:

$$\text{nth term} = 4 + 3n - 3$$

$$\text{nth term} = 3n + 1$$

Common trap: forgetting the $(n - 1)$

Students often write:

$$\text{nth term} = \text{first term} + n \times \text{difference}$$

That is usually wrong.
Why? Because the first term already exists before any extra jumps happen.
For the sequence

$$4,\ 7,\ 10,\ 13, \ldots$$

the 1st term should be 4.
Check:

$$4 + (1 - 1) \times 3 = 4$$

But if you use $4 + 1 \times 3$, you get 7, which is the second term, not the first.

Common trap: always check the difference is constant

The nth term formula only works for **arithmetic sequences** — ones where the difference between consecutive terms is the same every time.
For example, $1,\ 2,\ 4,\ 8,\ 16, \ldots$ does **not** have a constant difference. Applying first term $+ (n - 1) \times$ difference to this sequence will give wrong answers. Using the first gap of 1 gives $1 + (n - 1) \times 1 = n$, which predicts a 5th term of 5 — but the real 5th term is 16.
ThinkLab habit: before using the formula, always verify that the difference is constant by checking at least two consecutive gaps.

Worked Examples

Type A: Find a faraway term

Example 1: The 1000th term

Problem: What is the 1000th term in the sequence

$$4,\ 7,\ 10,\ 13,\ 16, \ldots$$

Step 1: Identify the first term and the difference
First term:

$$4$$

Difference:

$$3$$

Step 2: Use the nth term rule

$$\text{nth term} = 4 + (n - 1) \times 3$$

For the 1000th term:

$$4 + (1000 - 1) \times 3$$

$$4 + 999 \times 3$$

$$4 + 2997$$

$$3001$$

Answer: The 1000th term is **3001**.

Type B: Another arithmetic sequence

Example 2: The 100th term

Problem: What is the 100th term in the sequence

$$21,\ 52,\ 83,\ 114,\ 145, \ldots$$

Step 1: Find the common difference

$$52 - 21 = 31$$

So the sequence increases by 31 each time.
Step 2: Use the nth term rule

$$\text{nth term} = 21 + (n - 1) \times 31$$

For the 100th term:

$$21 + (100 - 1) \times 31$$

$$21 + 99 \times 31$$

$$21 + 3069$$

$$3090$$

Answer: The 100th term is **3090**.

Type C: Write the general rule

Example 3: Build the nth term

Problem: Find a rule for the sequence

$$5,\ 9,\ 13,\ 17, \ldots$$

Step 1: Identify the first term and difference
First term:

$$5$$

Difference:

$$4$$

Step 2: Write the nth term

$$\text{nth term} = 5 + (n - 1) \times 4$$

Step 3: Simplify

$$5 + 4n - 4$$

$$4n + 1$$

Answer: The nth term is

$$4n + 1.$$

Type D: A decreasing sequence

Example 4: Going down

Problem: Find the nth term of the sequence

$$20,\ 17,\ 14,\ 11, \ldots$$

Step 1: Identify the first term and the difference
First term: 20.
The sequence decreases, so the common difference is **negative**:

$$17 - 20 = -3$$

Step 2: Write the nth term

$$\text{nth term} = 20 + (n - 1) \times (-3)$$

Step 3: Expand carefully

$$= 20 - 3(n - 1)$$

$$= 20 - 3n + 3$$

$$= 23 - 3n$$

Check the first term: $23 - 3(1) = 23 - 3 = 20$ ✓
Check the second term: $23 - 3(2) = 23 - 6 = 17$ ✓
Answer: The nth term is $23 - 3n$.

Type E: Is a number in the sequence?

Example 5: Membership test

Problem: Is 202 a term in the sequence

$$4,\ 7,\ 10,\ 13,\ldots$$

Step 1: Write the nth term rule
From Example 1, the nth term is $3n+1$.
Step 2: Set it equal to 202

$$3n+1=202$$

$$3n=201$$

$$n=67$$

Since n is a whole number, 202 is in the sequence.
Answer: Yes, 202 is the **67**th term.

Type F: A growing visual pattern

Example 6: Square border pattern

Problem: A growing pattern uses dots arranged around the border of a square. The first three patterns look like this:

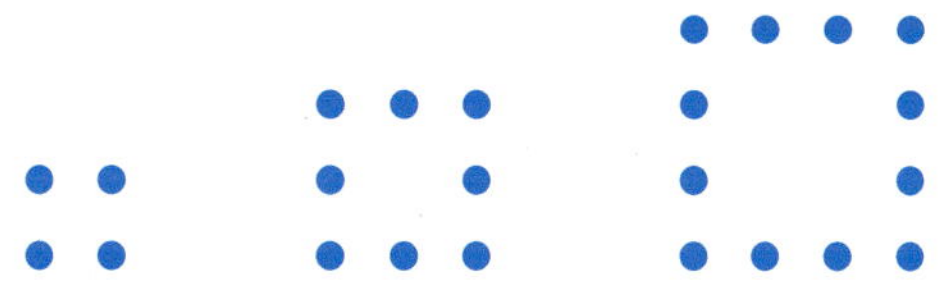
Pattern 1 Pattern 2 Pattern 3

How many dots will Pattern 20 use?
Step 1: Spot the pattern
Count the dots in each picture and write the sequence. Each new pattern adds one dot to each of the four sides, so the sequence is:

$$4,\ 8,\ 12,\ 16,\ldots$$

This starts at 4 and increases by 4 each time.
Step 2: Write the nth term rule

$$\text{nth term}=4+(n-1)\times 4$$

Step 3: Substitute $n=20$

$$4+(20-1)\times 4$$

$$4+19\times 4$$

$$4+76$$

80

Answer: Pattern 20 uses **80** dots.

ThinkLab habit: when a pattern grows geometrically, ask *where* the new dots appear. Here, four new dots arrive each time — one on each side — which is why the sequence jumps by 4.

Training Exercises

1. Find the next few terms

Write the next three terms in each sequence.

- a) 6, 10, 14, 18, ... ____________
- b) 15, 22, 29, 36, ... ____________
- c) 100, 95, 90, 85, ... ____________

2. Find the 20th term

Find the 20th term of:

- a) 4, 9, 14, 19, ...
- b) 12, 18, 24, 30, ...

3. Write the nth term

Find the nth term of each sequence.

- a) 3, 7, 11, 15, ...
- b) 10, 15, 20, 25, ...
- c) 7, 19, 31, 43, ...

4. Is it in the sequence?

For the sequence

$$5,\ 9,\ 13,\ 17,\ \ldots$$

decide whether each number is a term in the sequence.

- a) 81
- b) 82

Show your working.

5. The 100th term

What is the 100th term in the sequence

$$8,\ 15,\ 22,\ 29, \ldots$$

6. Growing pattern

A tile pattern grows like this:
Pattern 1 uses 9 tiles.
Pattern 2 uses 15 tiles.
Pattern 3 uses 21 tiles.
Pattern 4 uses 27 tiles.

- a) Write the nth term rule for the number of tiles.
- b) How many tiles will Pattern 30 use?

7. Decreasing sequence

Find the nth term of the sequence

$$50,\ 47,\ 44,\ 41, \ldots$$

Then decide whether 5 is a term in the sequence.

8. Reverse thinking

The *n*th term of a sequence is

$$4n + 3$$

- a) Write the first four terms.

- b) What is the 50th term?

9. Find the first term

The 1000th term of a sequence is 3997. The sequence is arithmetic and increases by 4 each time.
What is the first term?

Chapter Recap

What to remember from this chapter

- For an arithmetic sequence, find the first term and the common difference.
- Use

$$\text{nth term} = \text{first term} + (n - 1) \times \text{difference}$$

- Be careful not to forget the $(n - 1)$.
- A number belongs to the sequence only if the term number comes out as a whole number.
- Algebra helps you jump directly to faraway terms.

Before moving to Chapter 8

So far, algebra has helped us solve equations and describe patterns. In the next chapter, we use algebraic thinking to simplify arithmetic expressions quickly and spot elegant shortcuts in competition-style problems.

Quick Answer Check

Short answers

1. a) 22, 26, 30 b) 43, 50, 57 c) 80, 75, 70
2. a) 99 b) 126
3. a) $4n - 1$ b) $5n + 5$ c) $12n - 5$
4. a) yes b) no
5. 701
6. a) $6n + 3$ b) 183
7. $53 - 3n$; yes, it is the 16th term
8. a) 7, 11, 15, 19 b) 203
9. 1

Selected fuller solutions

Question 3(c)
The sequence is

$$7,\ 19,\ 31,\ 43, \ldots$$

First term: 7. Difference: $19 - 7 = 12$.
So the nth term is:

$$7 + (n - 1) \times 12 = 7 + 12n - 12 = 12n - 5.$$

Check: $12(1) - 5 = 7$ ✓ $12(2) - 5 = 19$ ✓

Question 4
The sequence

$$5,\ 9,\ 13,\ 17, \ldots$$

has first term 5 and difference 4, so:

$$\text{nth term} = 5 + (n - 1) \times 4$$

Simplify:

$$5 + 4n - 4$$

$$4n + 1$$

For 81:

$$4n + 1 = 81$$

$$4n = 80$$

$$n = 20$$

So 81 is in the sequence.
For 82:

$$4n + 1 = 82$$

$$4n = 81$$

$$n = 20.25$$

This is not a whole number, so 82 is not in the sequence.

Question 7
The sequence is

$$50,\ 47,\ 44,\ 41, \ldots$$

First term: 50.
Difference: -3.
So the nth term is $50 + (n - 1) \times (-3)$.
Simplify by expanding the bracket carefully:

$$50 - 3n + 3 = 53 - 3n$$

To check if 5 is a term in the sequence, set the nth term equal to 5:

$$53 - 3n = 5$$

Subtract 53 from both sides:

$$-3n = 5 - 53 = -48$$

Divide both sides by -3. A negative divided by a negative gives a positive result, so:

$$n = \frac{-48}{-3} = 16$$

Since n is a whole number, 5 is indeed a term in the sequence (it is the 16th term).

Question 9
The 1000th term is 3997 and the difference is 4.
Use:

$$\text{nth term} = \text{first term} + (n - 1) \times \text{difference}$$

So:

$$3997 = \text{first term} + (1000 - 1) \times 4 = \text{first term} + 3996$$

Subtract 3996:

$$\text{first term} = 1$$

Competition Algebra

Structure, Speed, and Smart Shortcuts

In earlier chapters, algebra helped us solve equations and describe patterns. In this chapter, algebra becomes a tool for **thinking faster**.

Many competition questions look like arithmetic, but the fastest solutions come from spotting algebraic structure:

- a common factor,
- a better grouping,
- a hidden friendly number,
- or a useful rearrangement.

What this chapter is really about

This chapter is not about introducing harder symbols. It is about seeing structure.
A strong competition student keeps asking:

- What is repeated?
- What can be grouped?
- Is there a common factor?
- Can I make 10, 100, or 1000?

The Main Idea

Concept 1: Distributive law in reverse

You already know that

$$a(b + c) = ab + ac.$$

In this chapter, we often use that idea backwards:

$$ab + ac = a(b + c).$$

This is called **factoring out a common factor**.
For example: $7 \times 13 + 7 \times 5$ has a common factor of 7, so:

$$7 \times 13 + 7 \times 5 = 7(13 + 5)$$

Then:

$$7(13 + 5) = 7 \times 18 = 126$$

Concept 2: Rearranging can make a problem easier

When a question has several terms, you do not always need to calculate from left to right.
Sometimes it is better to regroup terms to create a simpler expression.
For example:

$$25 + 48 + 75 + 52$$

can be grouped as:

$$(25 + 75) + (48 + 52)$$

which becomes:

$$100 + 100$$

$$200$$

The answer is the same, but the thinking is much faster.

Concept 3: Near-number thinking

Sometimes a multiplication is easy because one number is very close to a friendly number such as 10, 100, or 1000.
For example:

$$99 \times a = (100 - 1) \times a$$

So:

$$99 \times a = 100a - a$$

Similarly:

$$101 \times a = (100 + 1) \times a$$

$$101 \times a = 100a + a$$

This is often faster than long multiplication.

Common trap: calculating too early

A lot of students start long multiplication too soon.
For example, in

$$431 \times 236 + 764 \times 431$$

it is tempting to work out both products separately.
That is slow and error-prone.
A better question is:

What is common in both terms?

Since 431 appears in both, you should factor first.

Worked Examples

Type A: One common factor

Example 1: Factor first

Problem: Calculate

$$431 \times 236 + 764 \times 431$$

Step 1: Spot the common factor
Both terms contain: 431
So rewrite:

$$431 \times 236 + 764 \times 431 = 431 \times 236 + 431 \times 764$$

Step 2: Factor out 431

$$431(236 + 764)$$

Step 3: Add inside the bracket

$$236 + 764 = 1000$$

So:

$$431(236 + 764) = 431 \times 1000$$

$$431 \times 1000 = 431000$$

Answer: 431000.

Type B: More than one useful grouping

Example 2: Regroup cleverly

Problem: Calculate

$$143 \times 23 + 33 \times 31 + 44 \times 143 + 112 \times 33$$

Step 1: Look for common factors
Notice that the first and third terms both contain 143:

$$143 \times 23 + 44 \times 143$$

and the second and fourth terms both contain 33:

$$33 \times 31 + 112 \times 33$$

So regroup:

$$(143 \times 23 + 44 \times 143) + (33 \times 31 + 112 \times 33)$$

Step 2: Factor each group

$$143(23 + 44) + 33(31 + 112)$$

$$143 \times 67 + 33 \times 143$$

Step 3: Spot another common factor
Both terms now contain 143:

$$143 \times 67 + 33 \times 143 = 143 \times 67 + 143 \times 33$$

$$143(67 + 33)$$

Step 4: Add inside the bracket

$$67 + 33 = 100$$

So:

$$143(67 + 33) = 143 \times 100$$

$$143 \times 100 = 14300$$

Answer: 14300.

Type C: Friendly-number regrouping

Example 3: Make 100

Problem: Calculate

$$47 \times 18 + 53 \times 18$$

Step 1: Spot the common factor
Both terms contain:

$$18$$

So:

$$47 \times 18 + 53 \times 18 = (47 + 53) \times 18$$

Step 2: Add inside the bracket

$$47 + 53 = 100$$

So:

$$(47 + 53) \times 18 = 100 \times 18$$

$$100 \times 18 = 1800$$

Answer: 1800.

Type D: Arithmetic that is secretly algebra

Example 4: A hidden structure

Problem: Calculate

$$99 \times 37 + 37$$

Step 1: Notice the common factor
Both terms contain 37:

$$99 \times 37 + 37 = 99 \times 37 + 1 \times 37$$

$$99 \times 37 + 37 = (99 + 1) \times 37$$

Step 2: Simplify

$$(99 + 1) \times 37 = 100 \times 37$$

$$100 \times 37 = 3700$$

Answer: 3700.

Type E: Use a near number

Example 5: Use a near number

Problem: Calculate

$$99 \times 46$$

Step 1: Rewrite 99

$$99 = 100 - 1$$

So:

$$99 \times 46 = (100 - 1) \times 46$$

Step 2: Distribute

$$(100 - 1) \times 46 = 100 \times 46 - 1 \times 46$$

$$(100 - 1) \times 46 = 4600 - 46$$

Step 3: Simplify

$$4600 - 46 = 4554$$

Answer: 4554.

Type F: Decide which method is fastest

Example 6: Two ways, one better

Problem: Calculate

$$125 \times 16$$

You could multiply directly.
But a faster way is to notice:

$$125 \times 8 = 1000$$

So:

$$\begin{aligned} 125 \times 16 &= 125 \times (2 \times 8) \\ &= (125 \times 8) \times 2 \\ &= 1000 \times 2 = 2000. \end{aligned}$$

Answer: 2000.

Competition habit: do not just ask "Can I work it out?" Ask "What is the smartest way?"

Type G: Find the expression, not the variable

Concept 4: Read what is being asked

In many competition problems, you are **not** asked to find the value of the variable.
You are asked for the value of a related expression.
Sometimes you can answer directly without solving for the variable at all.
For example, if $3x + 5 = 17$, notice:

$$6x + 10 = 2(3x + 5) = 2 \times 17 = 34.$$

We found the value of $6x + 10$ without ever finding x.

Example 7: Use the equation directly

Problem: Given that $4n - 3 = 25$, find the value of $8n - 6$.
Method 1: Spot the structure
Notice that $8n - 6 = 2(4n - 3)$.
So:

$$8n - 6 = 2 \times 25 = 50.$$

Method 2: Solve first, then substitute
From $4n - 3 = 25$:

$$4n = 28, \text{ so } n = 7.$$

$$8n - 6 = 8 \times 7 - 6 = 56 - 6 = 50.$$

Both methods give 50. Method 1 is faster. In a timed competition, always check whether the expression you are asked for is a multiple of the given equation.
Answer: $8n - 6 = 50$.

Type H: Two expressions for the same thing

Concept 5: Two expressions for the same thing

When a problem gives you *two different descriptions of the same quantity*, write each description as an expression, then set them equal.
This turns a puzzle into an equation you can solve.
Competition habit: whenever you see a sentence like "A is also equal to B", or "the same number can be written as . . . ", reach for this move.

Example 8: Percentages that describe the same number

Problem: A number has the property that 20% of 30% of it is 3 less than 30% of 40% of the same number. Find the number.
Step 1: Write each description as an expression.
Let the number be x.
"20% of 30% of x" means "20% of (30% of x)". First take 30% of x, which is $0.3x$. Then take 20% of that:

$$0.2 \times 0.3x = 0.06x.$$

By the same method, 30% of 40% of x is:

$$0.3 \times 0.4 \times x = 0.12x.$$

Step 2: Use the "3 less than" relationship.
The first expression is 3 *less than* the second. So to make them equal, we subtract 3 from the bigger one:

$$0.06x = 0.12x - 3.$$

Step 3: Solve.
Add 3 to both sides:

$$0.06x + 3 = 0.12x.$$

Subtract $0.06x$ from both sides:

$$3 = 0.06x.$$

Divide both sides by 0.06:

$$x = 50.$$

Check: 20% of 30% of 50 is $0.06 \times 50 = 3$. 30% of 40% of 50 is $0.12 \times 50 = 6$. And 3 is indeed 3 less than 6. ✓
Answer: The number is 50.

ThinkLab habit: the hard part of this problem was not the arithmetic. It was noticing that the same x could be described two different ways.

Training Exercises

1. Find the expression

Given that $5x + 2 = 22$, find the value of:

- a) $10x + 4$
- b) $15x + 6$

2. Use the equation smartly

Given that $3a - 1 = 11$, find the value of $6a - 2$ without finding the value of a first.

3. Factor the common part

Calculate:

- a) $12 \times 7 + 12 \times 3$
- b) $45 \times 9 + 55 \times 9$
- c) $28 \times 13 + 28 \times 7$

4. Make a friendly number

Calculate:

- a) $39 \times 17 + 61 \times 17$
- b) $24 \times 50 + 76 \times 50$
- c) $88 + 37 + 12 + 63$

5. Reverse distributive law

Rewrite each expression in factored form.

- a) $7x + 7y$

- b) $5a + 15$
- c) $12n + 8$

6. Spot the common factor

Calculate:

$$273 \times 35 + 65 \times 273$$

7. Multi-step regrouping

Calculate:

$$27 \times 14 + 86 \times 27$$

8. Two unknowns from two clues

The sum of two numbers is 20 and their difference is 6. Find both numbers.
Hint: let the larger number be x. Then the smaller number is $x - 6$. Write one equation using the sum.

9. Hidden structure

Calculate:

$$99 \times 42 + 42$$

10. Near-number thinking

Calculate:

- a) 99×37
- b) 101×24

11. Heads and legs

On a farm there are some cows and some emus. Altogether there are 25 heads and 60 legs. How many cows and how many emus are there?
Hint: if there are x cows, how many emus are there? Then count the legs.

12. Paper strips

Two identical paper strips are glued together with an overlap in the middle to make one longer strip. When the overlap is 10 cm, the joined strip is 50 cm long.
If instead the two strips are joined with a different overlap to make a strip 56 cm long, how long is the new overlap?
Hint: first find the length of one strip.

13. Triplets and a younger brother

Three brothers are triplets (all the same age). Their fourth brother is 3 years younger than the triplets.
Which of these could be the total age of all four brothers: 53, 54, 56, 59, or 63?
Hint: let each triplet's age be x. Write an expression for the total of all four ages. What does the total have to be divisible by (plus or minus something)?

14. The fraction trick

Without calculating the numerator and denominator separately, find the value of

$$\frac{4444 + 5555}{1111 + 2222}.$$

Hint: look for a common factor in the top. Then look for the same factor in the bottom.

15. Four-term challenge

Calculate:

$$25 \times 18 + 75 \times 18 + 40 \times 7 + 60 \times 7$$

16. Two-layer factoring

Calculate:

$$36 \times 17 + 64 \times 17 + 36 \times 83 + 64 \times 83$$

Hint: factor out a common part from the first two terms and from the last two terms. Then look again at what you have.

17. Odd and even — algebraic proof

Any even number can be written as $2n$ where n is a whole number. Any odd number can be written as $2n + 1$.

- a) Write two different odd numbers using this notation. Call them $2m + 1$ and $2n + 1$ (use different letters for different numbers).
- b) Add them together and simplify. Is the result always even? Explain.
- c) What does this prove about the sum of any two odd numbers?

18. Explain the shortcut

A student writes:

$$64 \times 25 = 16 \times 100 = 1600.$$

Explain why this method works.

Chapter Recap

What to remember from this chapter

- Many arithmetic questions can be solved faster by spotting structure.
- Reverse distributive law helps factor out a common part.
- Regrouping terms can create friendly totals such as 10, 100, or 1000.
- Near-number thinking can turn a hard multiplication into an easy one.
- Avoid calculating too early.
- In competition maths, the smartest method is often better than the longest method.

Before moving to the final review chapter

You have now worked through the main algebra ideas in this book. The final chapter will mix these ideas together so that you can practise choosing the right strategy for each problem.

Quick Answer Check

Short answers

1. a) 44 b) 66
2. 22
3. a) 120 b) 900 c) 560
4. a) 1700 b) 5000 c) 200
5. a) $7(x+y)$ b) $5(a+3)$ c) $4(3n+2)$
6. 27300
7. 2700
8. 13 and 7
9. 4200
10. a) 3663 b) 2424
11. 5 cows and 20 emus
12. 4 cm
13. 53 years (the only total that fits)
14. 3
15. 2500
16. 10000
17. b) $(2m+1)+(2n+1) = 2m+2n+2 = 2(m+n+1)$, which is always even. c) The sum of any two odd numbers is always even.
18. Because
$$64 \times 25 = (16 \times 4) \times 25$$
and multiplication can be regrouped:
$$(16 \times 4) \times 25 = 16 \times (4 \times 25)$$
Now
$$4 \times 25 = 100$$
so
$$64 \times 25 = 16 \times 100 = 1600.$$

Selected fuller solutions

Question 6

$$273 \times 35 + 65 \times 273$$

Rewrite the second term:

$$273 \times 35 + 273 \times 65$$

Factor out 273:

$$273(35 + 65)$$

$$35 + 65 = 100$$

$$273 \times 100 = 27\,300$$

Question 11 (Heads and legs).
Let the number of cows be x. Then the number of emus is $25 - x$, because the heads add to 25.
Cows have 4 legs, emus have 2 legs. The total number of legs is:

$$4x + 2(25 - x) = 60$$

$$4x + 50 - 2x = 60$$

$$2x = 10$$

$$x = 5$$

So there are 5 cows and $25 - 5 = 20$ emus.
Check: $4 \times 5 + 2 \times 20 = 20 + 40 = 60$ legs. ✓

(If you chose emus as x instead, you would get $2x + 4(25 - x) = 60$*, which gives* $x = 20$*. Same answer: 20 emus and 5 cows. Either choice of variable works — the algebra just rearranges itself.)*

Question 12 (Paper strips).
Each strip has length L. When two strips overlap by 10 cm, the joined length is:

$$2L - 10 = 50$$

$$L = 30$$

So each strip is 30 cm long.
For the second case, let the overlap be v. Then:

$$2 \times 30 - v = 56$$

$$60 - v = 56$$

$$v = 4$$

Check: two 30 cm strips with a 4 cm overlap give $60 - 4 = 56$ cm. ✓
Answer: The overlap is 4 cm.

Question 13 (Triplets and a younger brother).
Let each triplet's age be x. The total age of all four brothers is:

$$3x + (x - 3) = 4x - 3$$

So the total must be of the form $4x - 3$. That means "total plus 3" must be divisible by 4.
Test each option:

- $53 + 3 = 56 = 4 \times 14$. Yes. Triplets would be 14, younger brother 11. ✓
- $54 + 3 = 57$. Not divisible by 4.
- $56 + 3 = 59$. Not divisible by 4.
- $59 + 3 = 62$. Not divisible by 4.
- $63 + 3 = 66$. Not divisible by 4.

Answer: The total could be 53.

ThinkLab note: this problem looks like a word problem but is really a divisibility problem in disguise. Algebra revealed the hidden pattern.

Question 14 (The fraction trick).
Look at the numerator:

$$4444 + 5555 = 1111 \times 4 + 1111 \times 5 = 1111 \times (4 + 5) = 1111 \times 9$$

Look at the denominator:

$$1111 + 2222 = 1111 \times 1 + 1111 \times 2 = 1111 \times (1 + 2) = 1111 \times 3$$

So the fraction becomes:

$$\frac{1111 \times 9}{1111 \times 3} = \frac{9}{3} = 3$$

Answer: The value is 3.

ThinkLab note: a student who tries to add $4444 + 5555$ *and* $1111 + 2222$ *directly can still get the right answer — but they will take much longer, and they will miss the point. The beauty of the problem is in the shared factor.*

Question 15

$$25 \times 18 + 75 \times 18 + 40 \times 7 + 60 \times 7$$

Group the first two terms:

$$25 \times 18 + 75 \times 18 = (25 + 75) \times 18$$

$$(25 + 75) \times 18 = 100 \times 18$$

$$100 \times 18 = 1800$$

Group the last two terms:

$$40 \times 7 + 60 \times 7 = (40 + 60) \times 7$$

$$(40 + 60) \times 7 = 100 \times 7$$

$$100 \times 7 = 700$$

Now add:

$$1800 + 700 = 2500$$

Question 16

$$36 \times 17 + 64 \times 17 + 36 \times 83 + 64 \times 83$$

Group the first two terms and the last two terms:

$$(36 + 64) \times 17 + (36 + 64) \times 83$$

$$100 \times 17 + 100 \times 83$$

Factor out 100:

$$100 \times (17 + 83)$$

$$100 \times 100 = 10\,000$$

Algebra Review and Challenge

Mixed Practice for Confident Problem Solvers

This final chapter brings together the main ideas from the book:

- visual algebra,
- expressions and equations,
- solving equations,
- simplifying,
- word problems,
- sequences,
- and competition-style shortcuts.

Some questions are routine, some are mixed, and some are designed to stretch your thinking. The goal is not only to get the right answer, but also to choose a good method and explain why it works.

How to use this chapter

Before starting a question, ask yourself:

- Is this an equation to solve?
- Is this a pattern to continue or generalise?
- Can I simplify first?
- Is there a faster structure hiding here?

That short pause often saves time.

Section A: Quick Warm-Up

1. Shape balance

If

$$\blacktriangle + \blacktriangle = 18,$$

find the value of ▲.

2. Expression or equation?

Write **Ex** for expression or **EQ** for equation.

- a) $3x + 5$ ________
- b) $2n + 4 = 18$ ________
- c) $7a$ ________
- d) $y - 9 = 12$ ________

3. Evaluate

If $x = 4$, find the value of $3x + 7$.

4. One-step equation

Solve:

$$x + 9 = 17$$

Section B: Core Algebra Practice

5. Two-step equation

Solve:

$$4x - 3 = 21$$

6. Fraction equation

Solve:

$$\frac{x}{5} + 2 = 9$$

7. Translate and solve

A number multiplied by 3, then increased by 4, gives 25.
Find the number.

8. Like terms

Simplify:

$$6a + 3b - 2a + 5b$$

9. Brackets

Simplify:

$$3(x + 4) - 2(x - 1)$$

10. A tempting wrong step

A student solves the equation

$$2x + 6 = 20$$

and writes the following working:

Step 1: Divide both sides by 2.

$$x + 6 = 10$$

Step 2: Subtract 6 from both sides.

$$x = 4$$

Something is wrong. The answer $x = 4$ does not satisfy the original equation.

- Which step went wrong, and why?
- Solve the equation correctly.

Section C: Problem Solving

11. Two people, one total

Mia and Noah have \$62 altogether. Noah has \$14 more than Mia.
How much money does Noah have?

12. Consecutive numbers

The sum of three consecutive numbers is 72.
Find the three numbers.

13. Perimeter problem

A rectangle has perimeter 50 cm. Its length is 3 cm more than its width.
Find the length.

14. Fraction word problem

A jar contains some marbles. One quarter of them are removed. Then 9 more marbles are removed. There are 27 marbles left.
How many marbles were in the jar at the start?

Section D: Patterns and Sequences

15. Continue the sequence

Write the next three terms:

$$11,\ 17,\ 23,\ 29, \ldots$$

16. Find the nth term

Find the nth term of the sequence

$$8,\ 13,\ 18,\ 23, \ldots$$

17. Faraway term

What is the 50th term of the sequence

$$6,\ 10,\ 14,\ 18, \ldots$$

18. Is it in the sequence?

Is 205 a term in the sequence

$$1,\ 5,\ 9,\ 13, \ldots$$

Show your working.

Section E: Competition Thinking

19. Factor first

Calculate:

$$37 \times 24 + 76 \times 37$$

20. Friendly-number regrouping

Calculate:

$$48 \times 25 + 52 \times 25$$

21. Four-term challenge

Calculate:

$$35 \times 16 + 65 \times 16 + 45 \times 12 + 55 \times 12$$

22. Factor to a round number

Calculate:

$$237 \times 43 + 57 \times 237$$

23. Find the expression

Given that $7x - 2 = 40$, find the value of $21x - 6$ without solving for x.
Hint: is $21x - 6$ a multiple of $7x - 2$?

Section F: Final Challenge Set

24. Choose your method

A farmer has some sheep and some ducks. Altogether, the animals have 38 heads and 100 legs. How many sheep and how many ducks does the farmer have?

- Solve this problem using algebra.
- Solve the same problem using a bar model, or by any other method you prefer.
- Which method felt faster? Which felt clearer? Write one sentence explaining your answer.

25. Two descriptions of the same number

A number has the property that half of it is 12 more than one third of it. What is the number?
Hint: write two expressions for the same number, then set them equal.

26. Find the expression, not the variable

Given that

$$5a - 7 = 23,$$

find the value of

$$15a - 21$$

without first finding the value of a.
Hint: compare $15a - 21$ with the left side of the given equation. Is it a multiple of that expression?

27. Consecutive numbers revisited

The sum of five consecutive whole numbers is 2025. What is the largest of the five numbers?
Hint: you can let the numbers be n, $n+1$, $n+2$, $n+3$, $n+4$, or you can let them be $n-2$, $n-1$, n, $n+1$, $n+2$. Which choice makes the algebra cleaner?

28. A classic disguise

Three friends share a collection of stickers. Ben has half as many as Ava. Chen has three times as many as Ava. Together they have 117 stickers.
How many stickers does Ava have?
Hint: draw a bar model. Let Ava be the variable; everyone else is described in terms of Ava. *That makes the total equation very short.*

Chapter Recap

What to remember from this book

- Algebra is about structure, not just letters.
- Choose your variable carefully.
- Keep equations balanced.
- Simplify step by step.
- Patterns can be described with general rules.
- In competition maths, the smartest method is often faster than direct calculation.

A final ThinkLab habit

When you meet a new problem, do not ask only:

What calculation should I do?

Also ask:

What structure do I notice?

That habit is one of the biggest differences between routine maths and strong problem solving.

Quick Answer Check

Short answers

1. ▲ = 9
2. a) Ex b) EQ c) Ex d) EQ
3. 19
4. 8
5. 6
6. 35
7. 7
8. $4a + 8b$
9. $x + 14$
10. Step 1 is wrong. Dividing both sides by 2 must be applied to *every term*, including the 6. The correct Step 1 gives $x + 3 = 10$, leading to $x = 7$. A cleaner first step is to subtract 6 before dividing.
11. Noah has $38
12. 23, 24, 25
13. 14 cm
14. 48 marbles
15. 35, 41, 47
16. $5n + 3$
17. 202
18. Yes, it is the 52nd term
19. 3700
20. 2500
21. 2800
22. 23700
23. 120
24. 12 sheep, 26 ducks
25. 72
26. 69
27. 407
28. 26

Selected fuller solutions

Question 9

$$3(x + 4) - 2(x - 1)$$

Expand the first bracket:

$$3(x + 4) = 3x + 12$$

Expand the second bracket:

$$-2(x - 1) = -2x + 2$$

Now combine:

$$3x + 12 - 2x + 2$$

Collect like terms:

$$(3x - 2x) + (12 + 2)$$

$$x + 14$$

Question 10 (Spot the mistake).
The student's Step 1 is the wrong move. When you divide both sides of an equation by 2, *every term* on both sides must be divided, including the constant on the left. The correct Step 1 would be:

$$\frac{2x+6}{2} = \frac{20}{2}$$

$$x+3 = 10$$

$$x = 7$$

Check: $2 \times 7 + 6 = 14 + 6 = 20$. ✓
A cleaner approach is to undo the addition first, then undo the multiplication:

$$2x+6 = 20$$

$$2x = 14$$

$$x = 7$$

ThinkLab lesson: when you divide both sides of an equation, the division applies to everything on each side — not just to the term that contains the variable.

Question 13
Let the width be w.
Then the length is $w+3$.
Perimeter:

$$2(\text{width} + \text{length}) = 50$$

So:

$$2(w + (w+3)) = 50$$

$$2(2w+3) = 50$$

$$4w+6 = 50$$

$$4w = 44$$

$$w = 11$$

Therefore, the length is $11 + 3 =$ **14** cm.

Question 14
Let the starting number of marbles be x.
After removing one quarter, the fraction remaining is $\frac{3}{4}$:

$$\frac{3}{4}x$$

After removing 9 more:

$$\frac{3}{4}x - 9$$

This equals 27:

$$\frac{3}{4}x - 9 = 27$$

Add 9 to both sides:

$$\frac{3}{4}x = 36$$

Multiply both sides by 4:

$$3x = 144$$

Divide both sides by 3:

$$x = 48$$

Quick check:

$$\frac{3}{4} \times 48 - 9 = 36 - 9 = 27\checkmark$$

Answer: There were 48 marbles at the start.

Question 18
The sequence

$$1,\ 5,\ 9,\ 13, \ldots$$

has first term 1 and common difference 4.
So the nth term is $1 + (n - 1) \times 4$.

$$1 + 4n - 4$$

$$4n - 3$$

Now test 205:

$$4n - 3 = 205$$

$$4n = 208$$

$$n = 52$$

Since n is a whole number, 205 is in the sequence.

Question 21

$$35 \times 16 + 65 \times 16 + 45 \times 12 + 55 \times 12$$

Group the first two terms:

$$35 \times 16 + 65 \times 16 = (35 + 65) \times 16$$

$$(35 + 65) \times 16 = 100 \times 16 = 1600$$

Group the last two terms:

$$45 \times 12 + 55 \times 12 = (45 + 55) \times 12$$

$$(45 + 55) \times 12 = 100 \times 12 = 1200$$

Add:

$$1600 + 1200 = 2800$$

Question 24 (Choose your method).
Algebra. Let the number of sheep be s. Then the number of ducks is $38 - s$ (because the heads total 38). Sheep have 4 legs, ducks have 2:

$$4s + 2(38 - s) = 100$$

$$4s + 76 - 2s = 100$$

$$2s = 24$$

$$s = 12$$

So there are 12 sheep and $38 - 12 = 26$ ducks.
Bar model / arithmetic method. If all 38 animals were ducks, there would be $2 \times 38 = 76$ legs. The actual total is 100 legs, which is $100 - 76 = 24$ extra legs. Each sheep contributes 2 extra legs compared with a duck. So the number of sheep is $24 \div 2 = 12$, and the number of ducks is 26.
Both methods give the same answer. The arithmetic method is often faster; the algebra method is more reliable when the numbers are awkward or when there are more than two types of animal.

Question 25 (Two descriptions of the same number).
Let the number be x.
Half of the number is $\frac{1}{2}x$. One third of the number is $\frac{1}{3}x$.
The problem says: half is 12 more than one third. So:

$$\tfrac{1}{2}x = \tfrac{1}{3}x + 12$$

Multiply both sides by 6 to clear the fractions:

$$3x = 2x + 72$$

$$x = 72$$

Check: half of 72 is 36. One third of 72 is 24. And 36 is indeed 12 more than 24. ✓

Question 26 (Find the expression, not the variable).
Notice that:

$$15a - 21 = 3(5a - 7)$$

So:

$$15a - 21 = 3 \times 23 = 69$$

ThinkLab note: always check whether the expression you are asked for is a simple multiple of the expression you are given. When it is, you can skip finding a altogether.

Question 27 (Consecutive numbers revisited).
The "middle number" method is far cleaner. Let the five numbers be

$$n-2,\ n-1,\ n,\ n+1,\ n+2.$$

When you add them, the -2 and $+2$ cancel, and so do the -1 and $+1$. The total is just $5n$:

$$5n = 2025$$

$$n = 405$$

The largest of the five numbers is $n+2 = 407$.
Using the alternative form $n, n+1, n+2, n+3, n+4$ also works, but the algebra is messier: you get $5n+10 = 2025$, then subtract 10, then divide by 5, then add 4 to n at the end.

ThinkLab lesson: the right variable choice can save several steps.

Question 28 (A classic disguise).
Let Ava have a stickers. Then Ben has $\frac{1}{2}a$ and Chen has $3a$.
Total:

$$a + \tfrac{1}{2}a + 3a = 117$$

$$\tfrac{2}{2}a + \tfrac{1}{2}a + \tfrac{6}{2}a = 117$$

$$\tfrac{9}{2}a = 117$$

$$a = 117 \times \tfrac{2}{9} = 26$$

So Ava has 26 stickers, Ben has 13, and Chen has 78. Check: $26+13+78 = 117$. ✓
Bar model view: let Ava be 2 blocks. Then Ben is 1 block (half of Ava) and Chen is 6 blocks (three times Ava). The three bars together hold $2+1+6 = 9$ blocks, which equal 117 stickers. So 1 block = 13, and Ava has 2 blocks = 26.
Notice that the bar model sidesteps the fraction $\frac{9}{2}$ completely. The block is just half of Ava's quantity.

Glossary

Arithmetic sequence — A sequence where the difference between consecutive terms is always the same. Example: 4, 7, 10, 13, ... has a common difference of 3.

Bar model — A picture made of equal rectangular blocks, used to show how quantities in a problem compare, share, or differ. Each block stands for one equal unit.

Coefficient — The number in front of a variable. In the expression $3x$, the coefficient is 3.

Common difference — The constant amount added (or subtracted) each time in an arithmetic sequence.

Common factor — A number or expression that divides evenly into two or more terms. In $8x + 12$, the common factor is 4, so the expression can be written as $4(2x + 3)$.

Consecutive numbers — Whole numbers that come one after another, such as 7, 8, 9.

Distributive law — The rule that $a(b + c) = ab + ac$. Used to expand brackets, and also in reverse to factor.

Equation — A mathematical statement with an equals sign, showing that two expressions have the same value. Example: $2n + 4 = 16$.

Expand — To multiply out brackets so that the expression is written without them. Example: $2(x + 3) = 2x + 6$. Expanding is the reverse of factoring.

Expression — A mathematical phrase without an equals sign. Example: $3n + 2$.

Factoring — Rewriting an expression as a product by taking out a common factor. Example: $6x + 9 = 3(2x + 3)$.

Inverse operation — The opposite of an operation, used to undo it. Addition and subtraction are inverses; so are multiplication and division.

Like terms — Terms with the same variable raised to the same power. $3x$ and $5x$ are like terms; $3x$ and $3x^2$ are not.

nth term — A rule that gives any term of a sequence directly from its position number n, without listing all the terms before it. For 4, 7, 10, 13, ... the nth term is $3n + 1$, so the 100th term is $3(100) + 1 = 301$.

Ratio — A way of comparing two or more quantities by how many equal parts each has. A ratio of 2 : 3 means two parts of the first for every three of the second.

Sequence — An ordered list of numbers that follows a rule.

Simplify — To rewrite an expression in a shorter or cleaner form without changing its value.

Substitute — To replace a variable with a number. If $x = 4$, then $3x + 2$ becomes $3(4) + 2 = 14$.

Variable — A letter that stands for a number. In algebra, a variable can represent an unknown to be found, or a quantity that can take many values.

About the Author

Tamim Shahriar is an educator, author, and the founder of Tamim's ThinkLab in Sydney. He holds a Bachelor's degree in Computer Science and Engineering and has over 20 years of experience across Bangladesh, Singapore, and Australia as a software engineer, engineering manager, and teacher.

Tamim is an academic counsellor for the Bangladesh Mathematical Olympiad and accompanied the Bangladesh national team to the 2014 International Mathematical Olympiad in Cape Town, South Africa. During his university years he was involved in competitive programming, and he now draws on both backgrounds in his teaching at ThinkLab, where he prepares students for mathematical problem solving and coding competitions.

His teaching focuses on critical thinking and creativity rather than memorisation—helping students understand why a method works, not just how to apply it.

Tamim has written several bestselling books on programming and mathematics. His most recent book, written for a global audience, teaches programming fundamentals through Python. ThinkLab Algebra is his first book for upper primary students written in English.

He lives in Sydney with his family.

Made in the USA
Las Vegas, NV
03 November 2026

57425293R00067